APPLETONS' MATHEMATICAL SERIES.

AN

ELEMENTARY ARITHMETIC.

BY

G. P. QUACKENBOS, A. M.,

AUTHOR OF

"AN ENGLISH GRAMMAR;" "FIRST LESSONS IN COMPOSITION;" "ADVANCED COURSE OF COMPOSITION AND RHETORIC;" "A NATURAL PHILOSOPHY;" "ILLUSTRATED SCHOOL HISTORY OF THE UNITED STATES;" "PRIMARY HISTORY OF THE UNITED STATES;" ETC.

UPON THE BASIS OF THE WORKS OF

GEO. R. PERKINS, LL.D.

NEW YORK:
D. APPLETON AND COMPANY,
443 & 445 BROADWAY.
1867.

PREFACE.

THIS volume is intended to follow our Primary Arithmetic, or that of any other series, or may be used as a first book with beginners that are not too young. It goes over the ground covered by the Primary, but in a style suited to minds somewhat more mature, enlarging on the subjects there treated, and introducing the pupil to many new ones. Besides the four fundamental operations, it gives a comprehensive view of Fractions, Federal Money, Reduction, and the Compound Rules, presenting under each a large collection of sums, in every variety, not too difficult, but so constructed as to require the pupil to think, and thus make the performance intelligent and not mechanical.

Convinced that too much theory and rule embarrass the young pupil, the author has in this respect sought to strike a happy mean,—presenting necessary explanations, but in few words; giving example sometimes the precedence over precept, and making rules intelligible by means of preliminary illustrations. Definitions are made brief and simple. Technical terms unnecessary at this stage of progress are avoided. The difficulties of beginners being appreciated, it is believed that they are here so met as to save the teacher the annoyance of constant demands for explanation.

In arrangement we trust some gain will be apparent; particularly in Compound Numbers, where, in stead of presenting the Tables in a body, to be confounded together in the pupil's mind, we immediately apply each Table, as soon as learned, in appropriate exercises, either mental or written. Attention is also invited to the inductive method used in developing the several subjects.

The teacher is requested to see that every principle is mastered as the pupil advances. A single defective link makes a whole chain worthless. If this suggestion is attended to, it is believed that the present work will make the young student thoroughly acquainted with the subjects it embraces, and properly prepare him for the next number of the series, THE PRACTICAL ARITHMETIC.

NEW YORK, August 6, 1863.

CONTENTS.

ELEMENTARY ARITHMETIC.

WHAT ARITHMETIC IS.

1. WE commence with ONE. We have one head, one mouth, one body.

One, a single thing, is called a **Unit.**

2. A unit joined to another unit, makes TWO. We have two eyes, two hands, two feet.

Another unit joined to two, makes THREE. Each of our fingers has three joints.

Another unit joined to three, makes FOUR.

So we may go on. Adding a unit each time, we get FIVE, SIX, SEVEN, EIGHT, NINE.

3. One, two, three, four, five, six, &c., are called **Numbers.**

A Number is, therefore, one unit or more.

4. **Arithmetic** treats of numbers.

5. Repeating the numbers in order—*one, two, three, four, five, six,* &c., is called **Counting.**

QUESTIONS.—1. With what do we commence? What is *one*, a single thing, called?—2. Of what is *two* made up? Of what is *three* made up? If we go on, adding a unit each time, what do we get?—3. What are *one, two, three, four,* &c., called? What is a Number?—4. Of what does Arithmetic treat?—5. What is Counting? Count nine. Count nine backwards—*nine, eight, seven,* &c.

NOTATION.

6. Every number has a name; as, *one*, *two*, *three*. In stead of writing out the name, however, we may represent it by a character; as, 1, 2, 3.

Notation is the art of expressing numbers by characters.

7. There are two systems of Notation, the Ar'abic and the Roman.

The Arabic Notation.

8. The Arabic Notation is so called because it was used by the Arabs. It employs these ten characters, called **Figures**:—

0	NAUGHT / CIPHER / ZERO	2	TWO	6	SIX
		3	THREE	7	SEVEN
		4	FOUR	8	EIGHT
1	ONE	5	FIVE	9	NINE

The first of these figures, 0, implies the absence of number. 0 cents means not a single cent.

9. The greatest number that can be expressed with one figure is nine. All the numbers above nine are expressed by combining two or more figures.

First, 1 is combined with each of the ten figures; then 2, forming the *twenties;* then 3, forming the *thirties;* then 4, forming the *forties*, &c.

---

6. How may numbers be represented? What is Notation?—7. How many systems of notation are there? Name them.—8. Why is the Arabic Notation so called? How many characters does it use? What are they called? Learn how to make the ten figures, and their names What does 0 imply?—9. What is the greatest number that can be expressed with one figure? How are all numbers above nine expressed?

10. The numbers formed of two figures are

10	ten	40	forty	70	seventy
11	eleven	41	forty-one	71	seventy-one
12	twelve	42	forty-two	72	seventy-two
13	thirteen	43	forty-three	73	seventy-three
14	fourteen	44	forty-four	74	seventy-four
15	fifteen	45	forty-five	75	seventy-five
16	sixteen	46	forty-six	76	seventy-six
17	seventeen	47	forty-seven	77	seventy-seven
18	eighteen	48	forty-eight	78	seventy-eight
19	nineteen	49	forty-nine	79	seventy-nine
20	twenty	50	fifty	80	eighty
21	twenty-one	51	fifty-one	81	eighty-one
22	twenty-two	52	fifty-two	82	eighty-two
23	twenty-three	53	fifty-three	83	eighty-three
24	twenty-four	54	fifty-four	84	eighty-four
25	twenty-five	55	fifty-five	85	eighty-five
26	twenty-six	56	fifty-six	86	eighty-six
27	twenty-seven	57	fifty-seven	87	eighty-seven
28	twenty-eight	58	fifty-eight	88	eighty-eight
29	twenty-nine	59	fifty-nine	89	eighty-nine
30	thirty	60	sixty	90	ninety
31	thirty-one	61	sixty-one	91	ninety-one
32	thirty-two	62	sixty-two	92	ninety-two
33	thirty-three	63	sixty-three	93	ninety-three
34	thirty-four	64	sixty-four	94	ninety-four
35	thirty-five	65	sixty-five	95	ninety-five
36	thirty-six	66	sixty-six	96	ninety-six
37	thirty-seven	67	sixty-seven	97	ninety-seven
38	thirty-eight	68	sixty-eight	98	ninety-eight
39	thirty-nine	69	sixty-nine	99	ninety-nine

EXERCISE.

Count from 1 to 99. Count from 99 to 1, backwards. With what figure do the thirties all begin? The sixties? Write the following numbers in figures:—thirty-seven; eleven; ninety-eight; eighty-nine; twelve; twenty; five; fifteen; fifty. What system of notation have you just used?

Units, Tens, Hundreds.

11. Ten, we have seen, is expressed thus—**10.** Then 1 in the second place denotes one *ten*. So, 2 in the second place (20) is two *tens*, &c.

Any figure standing in the second place represents so many *tens*. Hence it denotes ten times as much as if it stood in the first place.

12. The first place is called the **units' place.** The second is the **tens' place.**

13. Numbers greater than 99 must be expressed with more than two figures. A third place is thus required, which is called the **hundreds' place.**

A figure in the third place denotes ten times as much as if it stood in the second place, and a hundred times as much as if it stood in the first place.

14. To express the even hundreds, place the several figures in the third place, with naughts after them. Thus:—

100	one hundred	300	three hundred
200	two hundred	400	four hundred, &c.

15. Observe how the numbers between the even hundreds are expressed:—

101	one hundred and one	120	one hundred and twenty
102	one hundred and two, &c.	121	one hundred and twenty-one
110	one hundred and ten	201	two hundred and one
111	one hundred and eleven, &c.	202	two hundred and two, &c.

11. How is *ten* expressed? What does 1 in the second place denote? 2 in the second place? Any figure in the second place?—12. What is the first place called? What is the second place called?—13. How must numbers greater than 99 be expressed? What is the third place called? What value is denoted by a figure in the third place, compared with the second and the first?—14. How are the even hundreds expressed?—15. Learn how to express numbers between the even hundreds.

EXERCISE.

Count from 300 to 400. Count backwards from 900 to 800. Write on your slate the numbers from 500 to 600.

Write 3 units, 6 tens, 9 hundreds (963); 9 units, 6 tens, 3 hundreds; 7 tens, 5 units; 8 hundreds; 3 hundreds, 6 units; five hundreds, one ten; two tens.

Express in figures three hundred and ninety-six. Two hundred and twelve. Eighty-one. Four hundred and two. Eight hundred and thirty. Six hundred. Seventeen.

Thousands.

16. The greatest number that can be expressed with three figures, is 999. Next comes one thousand.

One thousand is expressed thus, 1000—by putting 1 in the fourth place, which is called the **thousands' place.**

17. The number of thousands is shown by the figure in the fourth place. Thus:—

2000	two thousand	6000	six thousand
3000	three thousand	7000	seven thousand
4000	four thousand	8000	eight thousand
5000	five thousand	9000	nine thousand

18. Ten thousand requires five figures to express it—10000. The fifth place is called that of **ten thousands.**

19. A hundred thousand requires six figures to express it—100000. The sixth place is called that of **hundred thousands.**

---

16. What is the greatest number that can be expressed with three figures? What comes next to 999? How is one thousand expressed? What is the fourth place called?—17. By what is the number of thousands shown? Give examples.—18. How many figures are required to express ten thousand? What is the fifth place called?—19. How many figures are required to express a hundred thousand? What is the sixth place called?

20. We have now had six places named :—units, tens, hundreds, thousands, ten thousands, hundred thousands.

These six places are divided into two Periods, of three figures each. The first Period is that of **units;** the second, that of **thousands.**

Hundred thousands	Ten thousands	Thousands	Hundreds	Tens	Units
6	5	4	3	2	1
Thousands			Units		

21. The second period is that of thousands. To express a given number of thousands, write the number in the second period. If there are no figures for the units' period, supply naughts.

EXAMPLES.—Write four hundred and twenty-three thousand. To do this, write four hundred and twenty-three, as already shown—423—for the second period. Supply naughts for the units' period, and we have the required number—423,000.

So we write seventeen thousand, 17,000.
Five hundred and one thousand, 501,000.
Six hundred and twenty thousand, 620,000.

If there are numbers corresponding to the places of the units' period, set them there in stead of naughts.

Forty-three thousand, two hundred and ninety, 43,290.
Seven thousand, one hundred and five, 7,105.
One hundred thousand, and sixty-seven. (As there are no hundreds in the units' period, supply 0.) 100,067.

---

20 Name the first six places in order. How are these six places divided? What is the first period called? What is the second period called? Name the places of the first period. Name those of the second period.—21. How are we to express a given number of thousands? What must be done, if there are no figures for the units' period? Learn how to write the examples given.

EXERCISE IN NOTATION.

Write the following numbers in figures:—

1. Five hundred and nine thousand.
2. Sixty-three thousand, two hundred and seven.
3. Eleven thousand, one hundred and eleven.
4. Seven hundred thousand and seventy.
5. Six thousand. Six hundred thousand.
6. Forty-three thousand and thirty-four.
7. Five hundred and twelve thousand, seven hundred.
8. Eighty thousand, eight hundred and eight.
9. Nine hundred and ninety-nine thousand.
10. Write the greatest number that can be expressed with three figures; with four figures; with five figures; with six figures.

Millions, Billions.

22. The third period is that of **millions.**

It consists of three places,—millions, ten millions, hundred millions.

Examples.—Two hundred million,	200,000,000.
Four hundred and one million,	401,000,000.
Seventy million, five hundred thousand,	70,500,000.
Six million, seventeen thousand, and seven,	6,017,007.

23. The fourth period is that of **billions.**

It consists of three places,—billions, ten billions, hundred billions.

Examples.—One hundred and two billion,	102,000,000,000.
Eleven billion, eleven thousand, and two,	11,000,011,002.
Four billion, twenty million, and six,	4,020,000,006.

---

22. What is the third period? Of how many places does it consist? Name them.—23. What is the fourth period? Of how many places does it consist? Name them.

Summing Up.

24. Name the periods in order, beginning at the right.

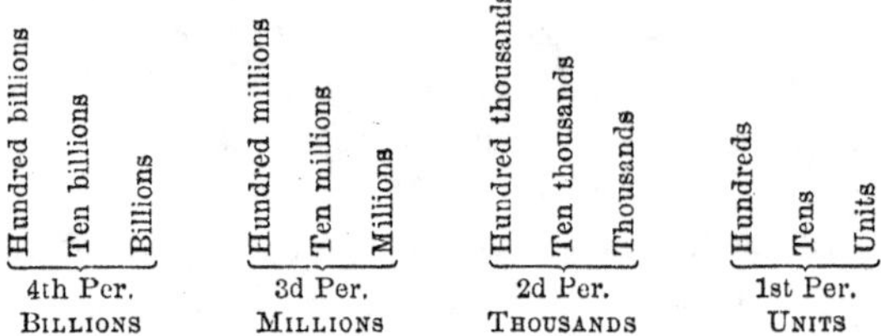

Name the places, beginning at the right.

10 units make 1 ten.
10 tens make 1 hundred.

Hence, removing a figure one place to the right, diminishes its value ten times; removing it one place to the left, increases it ten times.

25. Rule.—*Write billions in the fourth period, millions in the third, thousands in the second, units in the first, filling the vacant places with naughts, so as to have three places in each period.*

EXERCISE IN NOTATION.

Write the following numbers in figures, placing units under units, tens under tens, &c.:—

1. Four hundred and seventy-one billion, six thousand.
2. Ninety billion, three million, two thousand and four.
3. Eight hundred million, sixty thousand, one hundred.

---

24. What is the effect of removing a figure one place to the right? What is the effect of removing it one place to the left?—25. Give the rule for notation.

4. Six hundred and forty thousand, one hundred and one.
5. Nine million, fifty-seven thousand, and eight.
6. Eleven billion, forty-one million, two hundred and ten.
7. Thirty-six million, one hundred thousand, and twelve.
8. Ten billion, ten million, ten thousand, and three.
9. Six hundred and one million, two hundred thousand.
10. Eighty-nine thousand, three hundred and nineteen.
11. Twelve thousand, five hundred and eighty-seven.
12. Four hundred billion, four million, forty thousand.
13. Five million, eight hundred thousand, and seventy-five.
14. One billion, ten million, two hundred thousand, and six.
15. Fifty-seven million, three hundred and twenty-four.
16. Four million, two hundred and seventeen thousand, and fifty-eight.
17. Six hundred and nine billion, four hundred and sixty-six million, ninety-two thousand, three hundred and twenty-eight.

The Roman Notation.

26. The Roman Notation is so called because it was used by the ancient Romans.

It employs seven letters. I. denotes one; V., five; X., ten; L., fifty; C., one hundred; D., five hundred; M., one thousand.

27. These letters are combined to express numbers, according to the following principles:—

1. If a letter is repeated, its value is repeated. XX. is twenty; III. is three.

2. A letter of less value, placed after one of greater, unites its value to that of the latter. VI. is six.

---

26. Why is the Roman Notation so called? What does it use, to express numbers?—27. State the principles of the Roman Notation.

3. A letter of less value, placed before one of greater, takes its value from that of the latter. IV. is four.

4. A letter of less value, placed between two of greater, takes its value from that of the other two united. LIV. is fifty-four.

5. A bar over a letter increases its value a thousand times. $\overline{\text{V}}$. is five thousand.

TABLE.

I.	is	One.	L.	is	Fifty.
II.	"	Two.	LX.	"	Sixty.
III.	"	Three.	LXX.	"	Seventy.
IV.	"	Four.	LXXX.	"	Eighty.
V.	"	Five.	XC.	"	Ninety.
VI.	"	Six.	C.	"	One hundred.
VII.	"	Seven.	CI.	"	One hund. and one.
VIII.	"	Eight.	CIV.	"	One hund. and four.
IX.	"	Nine.	CX.	"	One hund. and ten.
X.	"	Ten.	CC.	"	Two hundred.
XI.	"	Eleven.	CCC.	"	Three hundred.
XII.	"	Twelve.	CCCC.	"	Four hundred.
XIII.	"	Thirteen.	D.	"	Five hundred.
XIV.	"	Fourteen.	DC.	"	Six hundred.
XV.	"	Fifteen.	DCC.	"	Seven hundred.
XVI.	"	Sixteen.	DCCC.	"	Eight hundred.
XVII.	"	Seventeen.	DCCCC.	"	Nine hundred.
XVIII.	"	Eighteen.	M.	"	One thousand.
XIX.	"	Nineteen.	MM.	"	Two thousand.
XX.	"	Twenty.	MMM.	"	Three thousand.
XXI.	"	Twenty-one.	MMMM.	"	Four thousand.
XXX.	"	Thirty.	$\overline{\text{V}}$.	"	Five thousand.
XL.	"	Forty.	$\overline{\text{X}}$.	"	Ten thousand.

What is the effect of placing a bar over a letter? How is five thousand denoted? Learn the Table.

28. We may, then, express numbers in three ways:—

1. With *words*, as is usual in printed books.

2. With *figures*, by the Arabic Notation, as in accounts and calculations.

3. With *letters*, by the Roman Notation, as in the headings of chapters.

EXERCISE IN NOTATION.

Write the following numbers first by the Arabic, and then by the Roman, Notation:—

1. Twelve.	6. One thousand.
2. Fifty-seven.	7. Ninety-nine.
3. Nine hundred.	8. Seven hundred.
4. Eighty-six.	9. Sixty-two.
5. Nineteen.	10. Four thousand.

11. Five thousand six hundred and seventy-three.
12. Three hundred and seventy-two.
13. Two thousand eight hundred and forty-one.
14. Nine thousand and twenty-seven.
15. Fifteen hundred and thirty-five.

Express the following numbers according to the Roman Notation: 12; 1,000; 749; 18; 203; 96; 660; 438; 29; 2,040; 85; 555; 10,801; 79; 5,002; 37; 394; 999; 2,062; 3,186; 119.

Express the following numbers according to the Arabic Notation: XII. LI. VIII. XLIII. XVI. LXXXIX. XCVIII. CCI. DXX. XXXIV. MD. IX. MCCXV. DCCCVII. XIV. MDCLXVI. $\overline{\text{V}}$.

x. vii. lv. cciv. xxxiii. xix. xlviii. xc. cxxi. xv. lxii.

---

28. How many ways are there of expressing numbers? What are they, and where is each used?

NUMERATION.

29. Numeration is the art of reading numbers expressed by figures.

30. In reading numbers, the following principles apply:—

1. We read by periods. Hence, if there are more than three figures, point off the number into periods of three figures each, beginning at the right.

2. Always begin to read at the left.

3. The right-hand figure and the right-hand period are never named as units, the word *units* being understood. We read 7 as *seven*, not *seven units*; 400 is read *four hundred*, not *four hundred units*.

4. Places containing 0 must be passed over in reading. We read 1062 *one thousand and sixty-two*, not *one thousand, no hundred, and sixty-two*.

31. RULE.—*Beginning at the right, point off the number into periods of three figures each.*

Beginning at the left, read the figures in each period as if they stood alone, adding the name of the period in every case except the last.

EXAMPLES.

10,709	Ten thousand, seven hundred and nine.
401,840	Four hundred and one thousand, eight hundred and forty.
6,023,070	Six million, twenty-three thousand, and seventy.

29. What is Numeration?—30. How do we read numbers? If there are more than three figures, what do we do? At which side do we begin to read? What is said of the right-hand figure and the right-hand period? What must be done in the case of places containing 0?—31. Give the rule for Numeration.

42,110,000	Forty-two million, one hundred and ten thousand.
870,025,002	Eight hundred and seventy million, twenty-five thousand, and two.
1,001,000,011	One billion, one million, and eleven.
19,056,007,000	Nineteen billion, fifty-six million, seven thousand.
123,400,789,000	One hundred and twenty-three billion, four hundred million, seven hundred and eighty-nine thousand.

1100 is read one thousand one hundred, or eleven hundred.
1200 " " one thousand two hundred, or twelve hundred, &c.

EXERCISE IN NUMERATION.

Read the following numbers:—

1.	903	15.	87123645	29.	MDCCCLXIII.
2.	8600	16.	476674983429	30.	$\overline{V}$CCCXCIX.
3.	21	17.	82000000117	31.	DCLXXXV.
4.	100075	18.	1413	32	XXXVIII.
5.	93282	19.	103600028	33.	MDXIX.
6.	3300000	20.	50500005050	34.	DCCXVII.
7.	463925	21.	442671376000	35.	MCXI.
8.	1650	22.	15000027	36.	LIX.
9.	1040400	23.	998899989898	37.	DXCVI.
10.	26308	24.	203013310031	38.	$\overline{L}$XIV.
11.	8005042	25.	410	39.	$\overline{X}$CCXXVI.
12.	741607	26.	5413760	40.	LXX.
13.	3821060	27.	83227	41.	MDCXVIII.
14.	600007	28.	14603000	42.	CCCCXLIV.

Review Questions.—What is a Unit? What is a Number? Of what does Arithmetic treat? What is Counting? What is Notation? Name the two systems of notation. What characters are used in the Arabic Notation? Name the periods in order, beginning at the right. Name the places. Give the rule for expressing numbers in figures. What characters are used in the Roman Notation? State the principles on which they are combined. What are used to express numbers, in making calculations? What, in accounts? What, in headings of chapters? What is Numeration? Give the rule for reading numbers.

ADDITION.

32. Two men are riding and three are walking; how many men are there in all?

Here we are required to find one number containing as many units as 2 and 3 together. This process is called Addition.

33. Addition is the process of uniting two or more numbers in one.

The one number thus obtained is called the **Sum.** 2 and 3 are 5; 5 is the sum.

Addition Table.

0 and 1 are 1; 0 and 2 are 2; 0 and any number make that number.

1 and	2 and	3 and	4 and	5 and
1 are 2	1 are 3	1 are 4	1 are 5	1 are 6
2 are 3	2 are 4	2 are 5	2 are 6	2 are 7
3 are 4	3 are 5	3 are 6	3 are 7	3 are 8
4 are 5	4 are 6	4 are 7	4 are 8	4 are 9
5 are 6	5 are 7	5 are 8	5 are 9	5 are 10
6 are 7	6 are 8	6 are 9	6 are 10	6 are 11
7 are 8	7 are 9	7 are 10	7 are 11	7 are 12
8 are 9	8 are 10	8 are 11	8 are 12	8 are 13
9 are 10	9 are 11	9 are 12	9 are 13	9 are 14
10 are 11	10 are 12	10 are 13	10 are 14	10 are 15
6 and	**7 and**	**8 and**	**9 and**	**10 and**
1 are 7	1 are 8	1 are 9	1 are 10	1 are 11
2 are 8	2 are 9	2 are 10	2 are 11	2 are 12
3 are 9	3 are 10	3 are 11	3 are 12	3 are 13
4 are 10	4 are 11	4 are 12	4 are 13	4 are 14
5 are 11	5 are 12	5 are 13	5 are 14	5 are 15
6 are 12	6 are 13	6 are 14	6 are 15	6 are 16
7 are 13	7 are 14	7 are 15	7 are 16	7 are 17
8 are 14	8 are 15	8 are 16	8 are 17	8 are 18
9 are 15	9 are 16	9 are 17	9 are 18	9 are 19
10 are 16	10 are 17	10 are 18	10 are 19	10 are 20

It is necessary to know the Tables perfectly, so as to say them backwards or forwards, out of order as well as in order. They must be mastered before going on.

34. Addition is denoted by an erect cross +, called **Plus**, placed between the numbers to be added. 6 + 5 is read *six plus five*, and means that *six and five are to be added.*

35. Two short horizontal lines =, placed between two quantities or sets of quantities, denote that they are equal. 6 + 5 = 11 is read *six plus five equals eleven*, and means that *the sum of six and five is eleven.*

36. Observe that if

3+2=5	4+5=9	3+7=10	5+8=13
then	then	then	then
13+2=15	34+5=39	53+7=60	25+8=33
23+2=25	44+5=49	63+7=70	45+8=53
33+2=35, &c.	54+5=59, &c.	73+7=80, &c.	85+8=93, &c.

37. Observe that 4 + 5 = 9 and 5 + 4 = 9.

Hence, when numbers are to be added, it makes no difference which is taken first.

EXERCISE ON THE ADDITION TABLE.

How many are 5 and 4? 4 and 5? 24 and 5? 25 and 4? 4 and 55? 94 and 5? 15 and 4? 3 and 2? 2 and 73?

How many are 7 and 1? 1 and 7? 67 and 1? 7 and 81? 11 and 7? 3 and 6? 6 and 3? 46 and 3? 6 and 23?

---

32. In the example given, what are we required to find? What is this process called?—33. What is Addition? What is the result of addition called? Recite the Table.—34 By what is addition denoted? What does 6+5 mean?—35. Describe the sign that denotes equality. What does 6+5 =11 mean?—36 If 3+2=5, then what follows? How much is 5+8? What follows?—37. How much is 4+5? How much is 5+4? What principle is laid down respecting numbers to be added?

How many are 6 and 2? 66 and 2? 2 and 86? 82 and 6? 2 and 6? 12 and 6? 32 and 6? 6 and 52? 16 and 2?

How many are 8 and 2? 2 and 8? 7 and 3? 3 and 7? 18 and 2? 97 and 3? 32 and 8? 78 and 2? 7 and 53?

How many are 3 and 5? 55 and 3? 3 and 95? 23 and 5? 13 and 5? 4 and 2? 64 and 2? 62 and 4? 4 and 72?

How many are 5 and 5? 2 and 7? 45 and 5? 52 and 7? 12 and 7? 37 and 2? 2 and 5? 42 and 5? 4 and 6?

How many are 4 and 3? 34 and 3? 93 and 4? 14 and 3? 4 and 4? 84 and 4? 4 and 54? 9 and 8? 4 and 9?

Find the sum of 1, 2, and 3. $8+2+3$. $4+1+5$. $6+3+1$. $3+4+2$. $6+3+3$. $4+6+7$. $3+7+10$. $20+3+4$.

Count by twos, commencing 2, 4, 6, 8, &c., up to 100.

Count by threes, commencing 3, 6, 9, 12, &c., up to 99.

Count by fours, commencing 4, 8, 12, 16, &c., to 100.

Count by fives, commencing 5, 10, 15, 20, &c., to 100.

MENTAL EXERCISES.

1. Seven metals were known to the ancients; 43 have been discovered since. How many metals are now known?

Ans. $7+43$ metals, or 50 metals.

2. How much will a boy earn in two weeks, if he earns 5 dollars the first week and 2 the second?

3. John Adams was president four years. He was 61 when he entered on the duties of the office; how old was he when he left it?

4. A gardener set out 9 trees one day, and 8 the next; how many did he set out both days?

5. If a house has 8 windows on one side, 6 on another, and 4 on a third, how many windows has it in all?

6. Napoleon had 4 brothers and 3 sisters, besides 5 that died in infancy; how many brothers and sisters had he in all?

7. How many pounds will a pair of chickens weigh, if each weighs three pounds?

8. My house is at the north end of a lake; Mr. A's is 3 miles south of the south end. If the lake is 5 miles long, how far is it from my house to Mr. A's?

9. A farmer who has 7 cows, buys 6 more; how many has he then?

10. In a jar that weighs 6 pounds, I put 11 pounds of butter; how much will the whole weigh?

11. A boy who bought a quire of paper for 20 cents, sold it so as to gain 5 cents; how much did he sell it for?

12. Washington was born in 1732. George III. was born 6 years later; what was the date of his birth?

13. A steamboat starts with 72 passengers. Three miles down the river, it receives 7 more passengers. How many has it then?

14. The Earth has 1 moon; Jupiter has 4 moons; Saturn, 8; Uranus, 6; Neptune, 1; how many moons does that make altogether?

38. Principles of Addition.

1. We must add things of the same kind. Therefore, in setting down numbers to be added, place units under units, tens under tens, &c.

2. The sum of units is units; of tens, tens; &c.

3. Always begin to add at the right.

4. Find the sum of each column; and, if it is expressed by one figure, set it down under the column added.

---

38. How must we set down numbers to be added? Why so? What is the sum of units? Of tens? Of hundreds? Where must we begin in adding? If the sum of each column is expressed by one figure, where must we set it?

EXAMPLE.—Add four million; three hundred and twenty-six thousand, two hundred and forty-seven; fifty-three thousand, four hundred and ten; and two hundred and twenty-one.

Operation.—Write down the numbers, units under units, tens under tens, &c.

Begin to add at the right.

1st column. 1 and 7 are 8. Set down 8.

2d. 2 and 1 are 3, and 4 is 7. Set down 7.	4000000
3d. 2 and 4 are 6, and 2 is 8. Set down 8.	326247
4th. 3 and 6 are 9. Set down 9.	53410
5th. 5 and 2 are 7. Set down 7.	221
6th. Bring down 3. *7th col.* Bring down 4. *Ans.*	4379878

Proof of Addition.

39. Proving an example is finding whether the work is correct.

40. Addition is proved by adding the columns from the top downward. If the sum is the same as when they are added from the bottom upward, we infer that the sum is right.

This Proof is based on the fact that, when numbers are to be added, it makes no difference in what order they are taken. The sum will be the same. If an error has been made in adding up, it is not likely to be repeated in adding down, and will thus be detected.

EXAMPLE.—Prove the above example. Add each column from the top downward. 7 and 1 are 8. 4 and 1 are 5, and 2 is 7. 2 and 4 are 6, and 2 is 8. 6 and 3 are 9. 2 and 5 are 7. Bring down 3. Bring down 4. *Answer*, 4379878 —the same as before. Hence the work is right.

4000000
326247
53410
221
4379878

Apply these principles in the example given.—39. What is meant by Proving an example?—40. How is addition proved? On what is this proof based? Prove the example just given.

EXAMPLES FOR THE SLATE.

41. Read the numbers added. Prove each example.

1. Add 600123, 154235, 34400, and 221. *Ans.* 788979.

2. Add 85026371, 41005, and 1810000. *Ans.* 86877376.

3. Find the value of 1234+455111+20234+12120.

4. Add 297661031851, 1135204022, 3115, and 1520010.

5. What is the sum of 201+100+2283+364114+322201?

6. Find the sum of thirty-one; one hundred and eleven; twenty thousand, four hundred and forty-two; seventeen thousand, one hundred and eleven; and sixty thousand, one hundred and three. *Ans.* 97798.

7. Add together seventeen million; one hundred and fifty thousand, one hundred; eight hundred and nine thousand, two hundred and seventy-two; and forty thousand, three hundred and sixteen. *Ans.* 17999688.

8. What is the sum of four million, eight hundred and twelve thousand, one hundred and two; thirty-one thousand, six hundred and twenty; one million, one thousand, and forty-five; and three thousand and thirty? *Ans.* 5847797.

9. Add together three hundred and twenty; one hundred and eleven million, two hundred and twelve; forty thousand, one hundred and thirty-two; and two million, one hundred and thirty-seven thousand. *Ans.* 113177664.

10. Find the sum of twenty billion, one thousand, and one; four thousand and eleven; four hundred and seven million, twenty thousand, six hundred and forty-two; and sixty-three thousand, one hundred and three. *Ans.* 20407088757.

11. A merchant sells $10000* worth of goods one day; $5123 worth, the next; and $2436 worth, the next. How much does he sell in all?

---

* This mark ($) denotes *dollars*. It is always placed before the number. $1000 is read *a thousand dollars*.

12. Three score are sixty. How many are three score and ten?

13. If I travel 1246 miles by steamer, 732 by railroad, and 21 by stage, how far do I travel altogether?

14. An army contains 23022 infantry privates; 710 infantry officers; 4000 cavalry, including officers; and 164 musicians. How many men in all in the army?

15. A. sells a vessel for $15420, which is $1355 less than it cost. What did the vessel cost?

16. The estate of a deceased man was divided as follows: his wife received two hundred and ten thousand dollars; his daughter, forty thousand two hundred and fifty dollars; his elder son, twenty-one thousand five hundred and six dollars; his younger son, twenty thousand one hundred and forty-two dollars. What was the value of the estate? *Ans.* $291898.

17. The planet Mars is 145,205,000 miles from the sun. Jupiter is 350,610,500 miles farther. How far is Jupiter?

Carrying.

42. The sum of a column may make more than one figure.

EXAMPLE.—Add 487 and 975.

```
 487
 975
----
1462
```

Begin at the right. 5 and 7 are 12—2 units and 1 ten. Set down 2 in the units' place, and add the 1 ten to the other tens.

1 and 7 are 8, and 8 is 16. 16 tens are 6 tens and 1 hundred. Set down 6 in the tens' place, and add the 1 hundred to the other hundreds.

1 and 9 are 10, and 4 is 14—14 hundreds, or 4 hundreds and 1 thousand. Answer, 1462.

---

42. With the given example, show what is meant by *carrying*. Give the rule for carrying.

43. This adding of the left-hand figure is called Carrying.

44. RULE FOR CARRYING.— *When the sum of a column is over* 9, *set down the right-hand figure, and carry the left-hand figure or figures to the next column.*

EXAMPLES FOR THE SLATE.

45. Read and add the following numbers. Prove each example.

(1)	(2)	(3)
24897	43345678	123423434559
64	1123355	23785432977
234567	7893	9876543696
2357911	54689	751002789
34567890	734321	10200859

4. Add 123405, 54210, 1794322, and 6541. *Ans.* 1978478.

5. Add 4275602, 45706, 5567801, and 365. *Ans.* 9889474.

6. Add 23, 6794, 896423, 597, and 16019. *Ans.* 919856.

7. What is the value of 965482190006+4063+8127299837+102009+9238675+67. *Ans.* 973618834657.

8. Find the number of days in a year, there being 31 days in January, 28 in February, 31 in March, 30 in April, 31 in May, 30 in June, 31 in July, 31 in August, 30 in September, 31 in October, 30 in November, and 31 in December.

9. The amount of treasure exported from California in 1861 was $40639089. This was $1664256 less than in 1860. What was it in 1860?

10. The two largest cities in Europe are London and Paris. The population of London in 1851 was 2362236; that of Paris, 1053262. What was the population of both?

11. Benjamin Franklin was born in 1706, and died at the age of 84. In what year did he die?

12. Rhode Island, the smallest state in the Union, contains 1306 square miles. Texas, the largest state, contains 236198 square miles more than Rhode Island. How many square miles in Texas?

13. In 1850, South Carolina produced 300901 bales of cotton; Alabama, 564429. How much did both produce?

14. The United States is made up of the Atlantic Slope, which contains 967576 square miles; the Mississippi Valley, which contains 1237311 square miles; and the Pacific Slope, which contains 778266 square miles. How many square miles in the whole United States?

15. In 1850, 4203064 copies of papers and periodicals were printed in Maine; 3067552, in New Hampshire; 2567662, in Vermont; 64820564, in Massachusetts; 2756950, in Rhode Island; 4267932, in Connecticut. How many were printed in all the New England states? *Ans.* 81683724.

46. Rule for Addition.

1. *Set units under units, tens under tens, &c.*

2. *Beginning at the right, find the sum of each column.*

3. *If the sum is expressed by one figure, write it under the column added; if not, set down the right-hand figure, and carry the left-hand figure or figures to the next column.*

EXERCISE.

47. The following examples are to be practised until they can be added at sight up and down, naming the results only. Thus in Example 1:—*five*, *eleven*, *thirteen*, *fifteen*, *eighteen*, *twenty-one*, *thirty*, *thirty-seven*—set down 7, and carry 3. *Three*, *seven*, *fourteen*, &c.

(1)	(2)	(3)	(4)
899697	891985	342687	185232
125429	263882	165431	965368
489843	885487	267994	905596
974583	298688	319416	78748
583162	656599	948668	588933
236972	677494	275773	879370
119876	681395	139915	899477
612345	684923	289076	276431

(5)	(6)	(7)	(8)
958576	758318	669786	895939
328492	272638	359628	765838
223967	364773	694279	678930
225523	525822	946335	514831
221679	294987	834569	455922
120739	162856	179145	379828
653865	175902	579757	263955
862781	943655	954852	675869
426893	682354	864351	416970
4022515	4181305	6082702	5048082

(9)	(10)	(11)
838725889	6158874991	652824777
189404376	8254889782	863928996
397846748	965449873	896956866
578757467	4833437997	885999586
949549259	2511486766	992845696
499699451	1896479693	894896986
887415762	65769958	885699577
739353343	4747786854	864712998
568239774	4598748765	893888579
659128135	2152759921	878994886
487816296	5439765653	893989466
18906147	8901729836	784985886
6814842647	50527180089	10389724299

SUBTRACTION.

48. Three boys are on the lawn. Two go into the house; how many are left?

Here we are required to take 2 from 3, or to find the difference between 2 and 3. This process is called Subtraction.

49. Subtraction is the process of taking one number from another.

50. The smaller number must always be taken from the greater. We can take 2 from 3, but not 3 from 2.

Subtraction Table.

0 from 1 leaves 1; 0 from 2 leaves 2; 0 from any number leaves that number.

1 from	2 from	3 from	4 from	5 from
1 leaves 0	2 leaves 0	3 leaves 0	4 leaves 0	5 leaves 0
2 leaves 1	3 leaves 1	4 leaves 1	5 leaves 1	6 leaves 1
3 leaves 2	4 leaves 2	5 leaves 2	6 leaves 2	7 leaves 2
4 leaves 3	5 leaves 3	6 leaves 3	7 leaves 3	8 leaves 3
5 leaves 4	6 leaves 4	7 leaves 4	8 leaves 4	9 leaves 4
6 leaves 5	7 leaves 5	8 leaves 5	9 leaves 5	10 leaves 5
7 leaves 6	8 leaves 6	9 leaves 6	10 leaves 6	11 leaves 6
8 leaves 7	9 leaves 7	10 leaves 7	11 leaves 7	12 leaves 7
9 leaves 8	10 leaves 8	11 leaves 8	12 leaves 8	13 leaves 8
10 leaves 9	11 leaves 9	12 leaves 9	13 leaves 9	14 leaves 9
6 from	7 from	8 from	9 from	10 from
6 leaves 0	7 leaves 0	8 leaves 0	9 leaves 0	10 leaves 0
7 leaves 1	8 leaves 1	9 leaves 1	10 leaves 1	11 leaves 1
8 leaves 2	9 leaves 2	10 leaves 2	11 leaves 2	12 leaves 2
9 leaves 3	10 leaves 3	11 leaves 3	12 leaves 3	13 leaves 3
10 leaves 4	11 leaves 4	12 leaves 4	13 leaves 4	14 leaves 4
11 leaves 5	12 leaves 5	13 leaves 5	14 leaves 5	15 leaves 5
12 leaves 6	13 leaves 6	14 leaves 6	15 leaves 6	16 leaves 6
13 leaves 7	14 leaves 7	15 leaves 7	16 leaves 7	17 leaves 7
14 leaves 8	15 leaves 8	16 leaves 8	17 leaves 8	18 leaves 8
15 leaves 9	16 leaves 9	17 leaves 9	18 leaves 9	19 leaves 9

51. The number to be subtracted, is called the **Subtrahend.** That from which the subtrahend is to be taken, is called the **Minuend.** The result, or what is left, is called the **Remainder.**

2 from 3 leaves 1; 2 is the subtrahend, 3 the minuend, 1 the remainder.

52. Subtraction is denoted by a short horizontal line —, called **Minus,** placed before the subtrahend.

3—2 is read *three minus two*, and means that 2 is to be subtracted from 3.

53. Observe that if

3—2=1	7—3=4	7—3=4	11—10=1
then	then	then	then
13—2=11	47—3=44	27—23=4	31—10=21
23—2=21	57—3=54	77—73=4	51—10=41
33—2=31, &c.	67—3=64, &c.	87—83=4, &c.	91—10=81, &c.

EXERCISE ON THE SUBTRACTION TABLE.

How many does 4 from 6 leave? 4 from 16? 4 from 36? 4 from 56? 14 from 16? 74 from 76? 24 from 26?

How much is 9—3? 29—3? 49—3? 69—3? 89—3? 9—6? 39—6? 59—6? 79—6? 99—6? 6—1? 11—1?

Take 2 from 4. 2 from 54. 2 from 94. 2 from 24. 3 from 4. 3 from 64. 3 from 34. 4 from 4. 4 from 94.

How much is 8—5? 78—5? 18—5? 48—5? 8—3? 28—3? 88—3? 68—3? 68—5? 5—5? 25—5? 35—5?

Subtract 3 from 5. 3 from 55. 33 from 35. 43 from 45. 2 from 5. 2 from 15. 12 from 15. 22 from 25. 2 from 85.

---

48. In the example given, what are we required to do? What is this process called?—49. What is Subtraction?—50. Which is the number to be subtracted? Recite the Table. What does 0 from 1 leave? 0 from 2? 0 from any number?—51. What is the number to be subtracted called? What is the number from which the subtrahend is to be taken called? What is the result called?—52. How is subtraction denoted?—53. How much is 3—2? What follows? How much is 11—10? What follows?

How many does 5 from 7 leave? 5 from 57? 5 from 87? 2 from 7? 2 from 17? 2 from 67? 4 from 7? 4 from 27?

How much is 8−4? 18−4? 38−34? 9−2? 69−2? 39−32? 49−42? 9−7? 19−17? 9−4? 59−4?

Take 9 from 10. 9 from 40. 9 from 60. 8 from 10. 8 from 16. 8 from 17. 7 from 11. 6 from 12. 5 from 10.

Count backward by twos from 100. Thus: 100, 98, &c.

Count backward by fives from 100. Thus: 100, 95, &c.

Count backward by tens from 100. Thus: 100, 90, &c.

Count backward by twos from 99. Thus: 99, 97, &c.

MENTAL EXERCISES.

1. Fifty metals are now known. Seven were known to the ancients. How many have been discovered since?

Ans. 50−7 metals, or 43 metals.

2. A boy buys a paper for 5 cents. He gives the newsman 10 cents. How much change will he get?

3. Ellen is 14 years old, and Jane is 6 years younger. How old is Jane?

4. A farmer who has 15 sheep, sells 7 of them. How many has he left?

5. A man is on his way home from a town twelve miles distant. When he has walked nine miles, how much farther has he to go?

6. A flower-girl who starts with 24 nosegays, comes back with 4. How many has she sold?

7. Five out of nine eggs turned out bad; how many were good?

8. If I buy a cow for $29, and sell her for $23, how much do I lose?

9. Leaving home with $15, I spend $6 and give $3 away. How much have I left?

54. Principles of Subtraction.

1. The smaller number is the one to be subtracted. Set it under the greater.

2. As we must subtract things of the same kind, place units under units, tens under tens, &c.

3. The difference between units and units is units; between tens and tens, tens; &c.

4. Begin to subtract at the right.

5. Take each figure of the subtrahend from the one above it, and set the remainder under the figure subtracted.

EXAMPLE.—From eight million six hundred and forty thousand nine hundred and fifty-seven, subtract two hundred and ten thousand four hundred and thirty-six.

Operation.—Write the smaller number under the greater, units under units, tens under tens, &c. Begin at the right.

6 from 7 leaves 1; set it down under the 6. 3 from 5 leaves 2. 4 from 9 leaves 5. 0 from 0 leaves 0. 1 from 4 leaves 3. 2 from 6 leaves 4. Bring down 8. Answer, 8430521.

```
8640957
 210436
-------
8430521
```

Proof of Subtraction.

55. Add the remainder and subtrahend. If their sum is equal to the minuend, the work is right.

EXAMPLE.—Prove the example just given. Add the remainder to the subtrahend. Their sum is 8640957, which is equal to the minuend. Therefore the work is right.

```
Rem. 8430521
Sub.  210436
     -------
Sum  8640957
```

54. How are the numbers to be set down in subtraction? Of what denomination is the difference between units and units? Between tens and tens? Where must we begin to subtract? How are we to proceed? Solve the example given.—55. How is subtraction proved? Prove the example just given.

EXAMPLES FOR THE SLATE.

56. Read the numbers given and the remainders. Prove each example.

1. From 1908647 take 2321.	6. 4678759−2678657.
2. From 897628 take 34527.	7. 579583099−20052064.
3. Take 1431 from 11463845.	8. 78199087−60152085.
4. From 673485 take 20432.	9. 3976189653−1042053.
5. Take 38143 from 1159463.	10. 12186947285−103014.

11. From six hundred and fifty-seven thousand eight hundred and forty-nine, subtract five hundred and twenty-one thousand six hundred and sixteen. *Ans.* 136233.

12. Take four million seven thousand one hundred and thirty-one, from forty-six million eighty-nine thousand five hundred and forty-two. *Ans.* 42082411.

13. The subtrahend is three million four hundred and forty-three thousand. The minuend is five million eight hundred and forty-nine thousand and six. What is the remainder? *Ans.* 2406006.

14. From eight hundred and forty-eight billion six hundred and ninety-seven million and ten, take thirty-one billion one hundred and forty-two million. *Ans.* 817555000010.

15. How much more is twelve billion eight hundred and seventy-nine million three hundred and sixty-four, than one billion two hundred and thirty-five million and twenty-two? *Ans.* 11644000342.

Carrying.

57. The lower figure may be greater than the one above it.

EXAMPLE.—From 738 take 419.

Begin at the right. We can not take 9 from 8, because 9 is greater than 8.

$$\begin{array}{r}738\\ \underline{419}\end{array}$$

Hence from the 3 tens we take 1 away, leaving 2 tens. The 1 ten taken away is equal to 10 units, which we add to the 8 units, making 18.

```
  2
 7̸38
 419
 ---
 319
```

Now we can subtract 9. 9 from 18 leaves 9; set it down. 1 from 2 (*not* 3) leaves 1; 4 from 7 leaves 3. Answer, 319.

To balance the 10 units added to the 8, we took away 1 of the tens from the upper line. But in stead of diminishing the upper figure 1, we may add 1 to the figure below it. This gives the same result, and, being more convenient, is the mode generally pursued.

Thus: 9 from 18, 9. 1 and 1 are 2; 2 from 3 leaves 1. 4 from 7, 3. Answer, 319.

```
 738
 419
 ---
 319
```

58. This adding of 1 to the lower figure is called **Carrying.**

59. We may have to carry several times in succession.

Example.—From 10000 take 9999. 9 from 10 leaves 1; set it down and carry 1. 1 and 9 are 10; 10 from 10 leaves nothing. Carry 1; 1 and 9 are 10; 10 from 10 leaves nothing. Carry 1; 1 and 9 are 10; 10 from 10 leaves nothing. Answer, 1.

```
 10000
  9999
 -----
     1
```

60. Rule for Carrying.—*When the lower figure is greater than the one above it, add* 10 *to the upper figure, subtract, and carry* 1 *to the next lower figure.*

---

57. Show, with the example given, how we proceed when the lower figure is greater than the one above it.—58. What is this adding of 1 to the lower figure called?—59. Show how we may have to carry several times in succession.—60. Give the rule for carrying.

EXAMPLES FOR THE SLATE.

61. Read the numbers. Subtract. Prove each example.

(1)	(2)	(3)
6310475	129060418801	600173240
3501768	68431160424	25823186

(4)	(5)	(6)
19372843	801009647258	415673285
9445276	7771819436	6906356

7. Take four thousand and thirteen, from one million and eleven. *Ans.* 995998.

8. From sixty-five thousand and seven, subtract nine hundred and ninety-nine. *Ans.* 64008.

9. From seven hundred and thirty-four thousand five hundred and twenty-one, subtract eighty-four thousand two hundred and eighty-three. *Ans.* 650238.

10. Subtract seven hundred and sixty-two million nine hundred thousand and seventy, from one billion fourteen thousand and nine. *Ans.* 237113939.

11. From two hundred and fifty-five thousand, take seven hundred and two. *Ans.* 254298.

62. Rule for Subtraction.

1. *Set the smaller number under the greater, units under units, tens under tens, &c.*

2. *Beginning at the right, take each figure of the subtrahend from the one above it, and set the remainder under the figure subtracted.*

3. *If the lower figure is greater than the one above it, add* 10 *to the upper figure, subtract, and carry* 1 *to the next lower figure.*

When a whole is given and one of its parts, what is the rule for finding the other part?

Subtract the given part from the whole.

When a whole is given and all its parts but one, what is the rule for finding that one?

Add the given parts, and subtract their sum from the whole.

When the year of a person's birth and that of his death are given, what is the rule for finding his age?

Subtract the earlier date from the later.

EXAMPLES FOR THE SLATE.

1. Howard, the philanthropist, was born in the year 1726. He died in 1790. To what age did he live?

2. Hudson explored the river called by his name in 1609. How long was this after the discovery of America, which took place in 1492?

3. How many years have elapsed from the discovery of America to the present time?

4. A man worth $47650, leaves $29855 to his wife and the rest to his son. What is the son's portion?

5. At an election 12572 votes were cast, of which the successful candidate received 7698. How many votes did the other candidate receive?

6. In the above election, what was the majority of the successful candidate—or, how many votes did he receive more than the other? *Ans.* 2824 votes.

7. Washington died in 1799, at the age of 67. In what year was he born?

8. In a state containing 2311786 inhabitants, there are 1143683 females; how many males are there?

Ans. 1168103 males.

9. A lady buys a house for $3000. She spends $169 on it for repairs, and then sells it for $3450. Does she gain or lose, and how much? *Ans.* Gains $281.

10. In 1854, 3038955 bushels of wheat were received in Chicago. In 1857, the receipts were 10554761 bushels. What was the increase in the 3 years? *Ans.* 7515806 bushels.

11. A house and lot are sold for $10575. If the lot cost $3250, and the house $8195, does the owner gain or lose, and how much? *Ans.* Loses $870.

12. The population of the city of New York in 1860 was 814287, which was 184477 more than it was in 1855. What was the population in 1855? *Ans.* 629810.

13. Two persons are 375 miles apart. They travel towards each other, one going 93 miles and the other 57. How far are they then apart? *Ans.* 225 miles.

14. A man owns a house valued at $6250, and stock to the amount of $4500. If he is in debt $3999, what is he worth in all? *Ans.* $6751.

15. If a merchant who has 19000 bushels of corn, sells one customer 4550 bushels, and another 398, how many bushels has he left? *Ans.* 14052 bushels.

16. A sells B 60 bushels of wheat for $74, a horse for $150, a wagon for $95, and $37 worth of butter. B pays $125 cash. How much does he still owe A? *Ans.* $231.

17. A farmer lays up 237 pounds of butter from his own dairy, and buys 349 pounds more of a neighbor. After giving away 50 pounds and selling 488, how much has he left? *Ans.* 48 pounds.

18. In a library of 1594 volumes, there are 727 French books, 586 German, and 138 Spanish. The rest are Italian; how many of the latter are there? *Ans.* 143 volumes.

19. From ten thousand subtract seventy-five. *Ans.* 9925.

MULTIPLICATION.

63. What will 4 pies cost, at 6 cents each?

If 1 pie costs 6 cents, 4 pies will cost 4 times 6 cents, or 24 cents. Here we are required to take 6 four times. This process is called Multiplication.

64. Multiplication is the process of taking a number a certain number of times.

Multiplication Table.

0 taken any number of times is 0. Once 0 is 0; twice 0 is 0; &c.
0 times any number is 0. 0 times 1 is 0; 0 times 2 is 0; &c.
Once any number is that number. Once 1 is 1; once 2 is 2; &c.

Twice	3 times	4 times	5 times	6 times	7 times
1 is 2	1 is 3	1 is 4	1 is 5	1 is 6	1 is 7
2 is 4	2 is 6	2 is 8	2 is 10	2 is 12	2 is 14
3 is 6	3 is 9	3 is 12	3 is 15	3 is 18	3 is 21
4 is 8	4 is 12	4 is 16	4 is 20	4 is 24	4 is 28
5 is 10	5 is 15	5 is 20	5 is 25	5 is 30	5 is 35
6 is 12	6 is 18	6 is 24	6 is 30	6 is 36	6 is 42
7 is 14	7 is 21	7 is 28	7 is 35	7 is 42	7 is 49
8 is 16	8 is 24	8 is 32	8 is 40	8 is 48	8 is 56
9 is 18	9 is 27	9 is 36	9 is 45	9 is 54	9 is 63
10 is 20	10 is 30	10 is 40	10 is 50	10 is 60	10 is 70
11 is 22	11 is 33	11 is 44	11 is 55	11 is 66	11 is 77
12 is 24	12 is 36	12 is 48	12 is 60	12 is 72	12 is 84

8 times	9 times	10 times	11 times	12 times
1 is 8	1 is 9	1 is 10	1 is 11	1 is 12
2 is 16	2 is 18	2 is 20	2 is 22	2 is 24
3 is 24	3 is 27	3 is 30	3 is 33	3 is 36
4 is 32	4 is 36	4 is 40	4 is 44	4 is 48
5 is 40	5 is 45	5 is 50	5 is 55	5 is 60
6 is 48	6 is 54	6 is 60	6 is 66	6 is 72
7 is 56	7 is 63	7 is 70	7 is 77	7 is 84
8 is 64	8 is 72	8 is 80	8 is 88	8 is 96
9 is 72	9 is 81	9 is 90	9 is 99	9 is 108
10 is 80	10 is 90	10 is 100	10 is 110	10 is 120
11 is 88	11 is 99	11 is 110	11 is 121	11 is 132
12 is 96	12 is 108	12 is 120	12 is 132	12 is 144

65. The number to be multiplied, is called the **Multiplicand.** That by which we are to multiply, is called the **Multiplier.** The result, or number obtained by multiplication, is called the **Product.**

3 times 2 is 6. 2 is the multiplicand, 3 the multiplier, 6 the product.

66. The multiplicand and multiplier are called **Factors** of the product. 2 and 3 are factors of 6.

67. Multiplication is denoted by a slanting cross $\times$, placed between the factors. 2×3 is read, and denotes, *two multiplied by three.*

68. The multiplier shows how many times the multiplicand is to be taken. Multiplying 2 by 3 is taking 2 three times: $2+2+2=6$. $2\times3=6$.

Multiplying is therefore a short way of adding a number to itself.

How many trees are there in three rows of four trees each?

We may do this sum in two ways. We may say, If one row contains 4 trees, 3 rows will contain 3 times 4 trees, or 12 trees. This is doing it by multiplication.

Or we may say, As there are 4 trees in each row, in the three rows there are $4+4+4$ trees, or 12 trees. This is doing it by addition. The result is the same.

---

63. Repeat the example given. What are we here required to do? What is this process called?—64. What is Multiplication? How much is 0, taken any number of times? How much is 0 times any number? How much is once any number? Recite the Table.—65. What is the number to be multiplied called? What is that by which we are to multiply called? What is the result obtained by multiplication called?—66. What are the multiplicand and multiplier called?—67. How is multiplication denoted?—68. What does the multiplier show? Multiplying is a short way of doing what? Illustrate this.

69. When two numbers are to be multiplied together, it makes no difference in the result which is taken as the multiplicand, and which as the multiplier. $4 \times 3 = 12$. $3 \times 4 = 12$.

We have 12 trees in the above engraving, whether we take them crosswise as forming 3 rows of 4 each, or lengthwise as forming 4 rows of 3 each.

70. The product is of the same kind as the multiplicand. 3 times 2 *men* is 6 *men;* 3 times 2 *apples* is 6 *apples;* &c.

MENTAL EXERCISES.

1. If there are seven days in a week, how many days are there in nine weeks?

MODEL.—If there are 7 days in 1 week, in 9 weeks there are 9 times 7 days, or 63 days. Answer, 63 days.

2. If a man's wages are three dollars a day, how much will he earn in five days? How much in ten days?

3. If a stage makes 6 trips a day, of 3 miles each, how many trips will it make in a week, excluding Sunday?

4. What is the cost of 11 tons of coal, at $6 a ton?

5. How long will it take one boy to do a job which it takes five boys eight days to do?

6. At the rate of 4 for a cent, how many crackers can be bought for 7 cents? How many for 12 cents?

7. Four quarts make a gallon. How many quarts will a five-gallon jug hold?

8. What will 12 tables cost, at $11 apiece?

9. If an omnibus carries 12 passengers each trip, how many does it carry in 6 trips? In 8 trips?

---

69. In multiplying two numbers, what is found to make no difference? Illustrate this with the engraving.—70. Of what kind is the product?

10. If a box of tea lasts 8 persons 11 weeks, how long will it last 1 person at the same rate?

11. There are 2 pints in a quart. How many pints in 3 quarts? In 5 quarts? In 7 quarts? In 9 quarts?

12. If two boys do 4 sums apiece every day, how many will both do in 3 days?

13. Three fields contain 3 trees each. Under each tree 3 cows are lying. How many cows are in the three fields?

14. If 2 apples can be bought for 1 cent, how many can be bought for 10 cents?

15. Four girls have 2 hens each, and each hen has 6 chickens. How many chickens have they altogether?

EXAMPLES FOR THE SLATE.

71. Multiply 60123 by 3.

Here the multiplier is a single figure. Set it under the units' figure of the multiplicand. Beginning at the right, multiply each figure of the multiplicand by the multiplier, setting each product in a column with the figure multiplied.

60123
3
180369

Three times 3 is 9; 3 times 2 is 6; 3 times 1 is 3; 3 times 0 is 0; 3 times 6 is 18. The last product consists of two figures. Set it down with its right-hand figure under the figure multiplied. Answer, 180369.

	(1)	(2)	(3)	(4)
Multiply	9432	71232	52122	81010
By	2	3	4	5

5. At 7 cents a pound, what will 1010 pounds of cheese cost?

6. If a man lays up $510 a year, how many dollars will he be worth in 9 years? *Ans.* $4590.

Carrying.

72. When a figure is multiplied, the product may consist of two figures. As we can not set them both down under the figure multiplied, we place the right-hand figure there, and add the other figure to the next product.

This adding of the left-hand figure to the next product is called **Carrying**.

Example.—Multiply 7608 by 7.

```
 7608
    7
-----
53256
```

Begin at the right. 7 times 8 is 56,—5 tens and 6 units. Set the 6 units in the units' place, and carry the 5 tens to the next product.

7 times 0 tens are 0 tens, and 5 makes 5 tens. Set it down.

7 times 6 hundreds are 42 hundreds,—or, 4 thousands and 2 hundreds. Set the 2 hundreds in the hundreds' place, and carry the 4 thousands to the next product.

7 times 7 thousands are 49 thousands, and 4 makes 53. This being the last product, set down both figures. Answer, 53256.

73. Rule for Carrying.—*When any product exceeds 9, set down the right-hand figure in the same column with the figure multiplied, and add the remaining figure or figures to the next product.*

EXAMPLES FOR THE SLATE.

1. 23618 × 6. *Ans.* 141708.	5. 28946735 × 9.
2. 41726 × 7. *Ans.* 292082.	6. 9156738 × 8.
3. 683918547 × 4.	7. 4507956 × 6.
4. 9254658973 × 5.	8. 51708949326 × 8.

72. When, on multiplying a figure, we get a product of two figures, what must we do? What is this process called? Illustrate the mode of carrying, with the example given.—73. Recite the rule for carrying in multiplication.

9. The multiplicand is ninety-three thousand and forty-six; the multiplier is 3; what is the product? *Ans.* 279138.

10. How much is nine times seven hundred and fifty-eight thousand and twenty-nine? *Ans.* 6822261.

11. There are 365 days in a year. How many days are there in three years?

12. The moon is 240000 miles from the earth. If there were a comet 6 times that distance from the earth, how far would it be? *Ans.* 1440000 miles.

13. Nine million eight hundred thousand two hundred and fifty-seven is one factor; seven is the other; what is the product? *Ans.* 68601799.

Multiplying by two or more figures.

74. Multiply by 10, 11, and 12, in one line.

EXAMPLE.—Multiply 541 by 12.

12 times 1 is 12; set down 2 and carry 1. 12 times 4 is 48, and 1 makes 49; set down 9 and carry 4. 12 times 5 is 60, and 4 makes 64. Set it down. Answer, 6492.

$$\begin{array}{r} 541 \\ 12 \\ \hline 6492 \end{array}$$

75. When the multiplier is over 12, multiply by its figures separately.

EXAMPLE.—Multiply 287 by 156.

We can not multiply by 156 at once. Hence we first multiply by the 6 units; then by the 5 tens, or 50; then by the 1 hundred. Thus we get three **Partial Products,** as they are called; and, adding these, we get the whole product.

---

74. How are we to multiply by 10, 11, and 12? Give an example.—75. When the multiplier is over 12, how are we to proceed? What are the results obtained by multiplying by each figure called?

Multiplicand	287	
Multiplier	156	
Partial Products	1722	$= 287 \times 6$
	1435	$= 287 \times 50$
	287	$= 287 \times 100$
Product	44772	$= 287 \times 156$

76. General Rule.

1. *Set the multiplier under the multiplicand, units under units, tens under tens, &c.*

2. *Beginning at the right, multiply by each figure of the multiplier in turn, setting the first figure of each partial product in a column with the figure used in multiplying.*

3. *Add the partial products, and their sum will be the whole product.*

Proof.

77. Multiply the multiplier by the multiplicand. If this product is the same as the former one, the work is right.

EXAMPLE.—Prove the example given above.

Multiply the multiplier 156, by the multiplicand 287. The product is 44772, which is the same as the former product. Hence the work is right.

```
  156
  287
 ----
 1092
1248
312
-----
44772
```

78. When two numbers are to be multiplied together, it is usual to take the one with the fewer figures for the multiplier.

Multiply 287 by 156.—76. Recite the general rule for multiplication.—77. How is multiplication proved? Prove the example last given.—78. Which of two numbers is it usual to take for the multiplier?

EXAMPLES FOR THE SLATE.

Find the value of the following. Prove each example.

1. 356 × 11.	5. 1678 × 948.
2. 289 × 12.	6. 3548 × 751.
3. 165 × 234.	7. 2674 × 863.
4. 329 × 576.	8. 9463 × 594.

9. How much is 42 times 6257? *Ans.* 262794.
10. How much is 53 times 8164? *Ans.* 432692.
11. Multiply 4567031 by 147. *Ans.* 671353557.
12. Multiply 4905604 by 263. *Ans.* 1290173852.
13. Multiply 65728913 by 5674. *Ans.* 372945852362.
14. Multiply 8549412 by 32895. *Ans.* 281232907740.
15. Multiply 5076398 by 9168. *Ans.* 46540416864.
16. Multiply 765987 by 7896. *Ans.* 6048233352.
17. Multiply 542896 by 6892. *Ans.* 3741639232.
18. Multiply 99887 by 2673. *Ans.* 266997951.
19. Find the product of six thousand four hundred and twelve, and seventy-five thousand eight hundred and thirty-nine. *Ans.* 486279668.
20. Multiply eighty thousand four hundred and sixty-seven, by nine hundred and seventy-six. *Ans.* 78535792.
21. In a certain orchard, there are 15 rows of trees, and 19 trees in each row. How many apples will the whole orchard produce, if the average yield of each tree is 948 apples? *Ans.* 270180 apples.

0 in the multiplier.

79. *When 0 occurs in the multiplier, bring it down, and go on multiplying by the next figure, all in the same line.*

---

79. When 0 occurs in the multiplier, how are we to proceed?

EXAMPLE.—Multiply 2473 by 5008.

Beginning at the right, multiply by 8. Then bring down the first 0 of the multiplier in the tens' place, and the second 0 in the hundreds' place. Then multiply by 5, placing the product in the same line. Finally, add the partial products. Ans., 12384784.

```
    2473
    5008
--------
   19784
 1236500
--------
12384784
```

EXAMPLES FOR THE SLATE.

1. Multiply 46893 by 40308. *Ans.* 1890163044.

2. Multiply 962734 by 700906. *Ans.* 674786037004.

3. Multiply eighty thousand and twenty-nine by five thousand and seven. *Ans.* 400705203.

4. The Morris and Essex Canal, which is 101 miles in length, cost on an average $30693 a mile. What was the whole cost? *Ans.* $3099993.

5. If 19008 pounds of hay are required for the horses of a cavalry regiment one day, how many pounds will be needed for 206 days?

6. What would be the cost of constructing 309 miles of plank road, at $3975 a mile?

7. How many apples will an orchard containing 208 trees produce, if the average yield is 1269 apples for each tree?

8. In 3 editions of 750 books each, how many pages, if each book contains 407 pages? *Ans.* 915750 pages.

First find how many books there are, then how many pages.

Naughts at the right.

80. *When there are naughts at the right of either factor or both, multiply the other figures, and annex to their product as many naughts as are at the right of both factors.*

Solve the example given above.—80. When there are naughts at the right of either factor or both, how are we to proceed?

EXAMPLE.—Multiply 450000 by 370.

Cutting off the naughts, multiply 45 by 37. Then, as there are four naughts at the right of the multiplicand and one at the right of the multiplier, annex five naughts to the product. Answer, 166500000.

```
 45|0000
 37|0
 ---
 315
135
---------
166500000
```

81. Multiplying a number by 1 does not affect its value. Hence, *to multiply by* 10, 100, 1000, *&c., simply annex as many naughts as are in the multiplier.*

37×10=370   37×100=3700   37×1000=37000

EXAMPLES FOR THE SLATE.

Find the value of the following:—

1. 6706×10.
2. 89271×100.
3. 50860×120.
4. 7800×4300.
5. 867×10000.
6. 8000×700.
7. 290000×98.
8. 568×11000.
9. 74600000×56.
10. 6060×7040.

11. If there are 100 cents in one dollar, how many cents are there in $60400? *Ans.* 6040000 cents.

12. How many men were there in 11 Roman legions, if there were 4500 men in one legion?

13. 100 pounds make a hundred-weight. How many pounds in a ton, which contains 20 hundred-weight?

14. The Roman soldiers on a march carried 60 pounds' weight apiece. How many pounds did 3000 soldiers carry?

15. What will five hundred acres of land cost, at thirty dollars an acre? *Ans.* $15000.

16. What will 1000 acres cost, at $20 an acre?

Multiply 450000 by 370.—81. What is the effect of multiplying a number by 1? To multiply by 10, 100, 1000, &c., what are we to do?

Multiplying by factors.

82. When the multiplier is itself a product, we may multiply either by the whole, or by its factors in turn. The result will be the same.

EXAMPLE.—Multiply 84 by 36.

$36 = 6 \times 6$ or, 9×4 or, 12×3.

84	84	84	84
36	6	9	12
504	504	756	1008
252	6	4	3
3024	3024	3024	3024

Whether we multiply by 36 at once or by the different sets of factors that produce it, we get the same product,—3024. Multiplying by factors, therefore, proves whether the product first obtained is right.

EXAMPLES FOR THE SLATE.

In these examples, first multiply by the multiplier at once. Then prove the result by multiplying by its factors.

1. 875×81 (9×9).
2. 5761×44 (11×4).
3. 436×72 $(6 \times 12$ or $8 \times 9)$.
4. 9000×27.
5. 18274×24.
6. 45000×20.

7. How many ounces in 253 pounds of sugar, if there are 16 ounces in one pound?

8. How many bricks are there in 18 loads, if each load contains 1250 bricks?

9. What will 28 horses cost, at $75 apiece?

10. How many pounds in 32 firkins of butter, allowing 56 pounds to the firkin?

---

82. When the multiplier is itself a product, what two modes of proceeding are there? Illustrate these, with the given example.

MISCELLANEOUS EXAMPLES.

1. The sun is 95000000 miles from the earth; the moon is 240000 miles. How much farther from us is the sun than the moon? *Ans.* 94760000 miles.

2. A man worth $10000 gains $1000 by one sale, and loses $500 by another. What is he then worth? *Ans.* $10500.

3. If the railroad from Albany to Buffalo, 326 miles in length, cost $25649 a mile, what was the whole cost? *Ans.* $8361574.

4. How far will a locomotive travel in 7 days of 24 hours each, if it goes 30 miles an hour? *Ans.* 5040 miles.

5. A man left each of his 3 daughters 510 acres of land, and his wife 100 acres more than all his daughters together. What was the wife's portion? *Ans.* 1630 acres.

6. A farmer raised 1570 bushels of potatoes, and bought 730 bushels more. After selling 4 lots of 500 bushels each, how many bushels had he on hand? *Ans.* 300 bushels.

7. How many hours in 365 days of 24 hours? *Ans.* 8760.

8. The earth moves in its orbit 68000 miles an hour; how far will it move in 365 days? *Ans.* 595680000 miles.

9. The President's salary is $25000. If he spends $19500 a year, how much will he save during his 4 years' term?

How much will he save in 1 year? How much, then, in 4 years?

Ans. $22000.

10. A has 1060 sheep; B has 849; C has 1276. How many must C buy, to have as many as A and B together?

Find how many A and B have. Then find the difference between this number and C's.

Ans. 633 sheep.

11. Daniel Webster was born in 1782 and died in 1852. How old was he?

12. At $3 a yard, what will 101 pieces of cloth cost, if each piece contains 40 yards? *Ans.* $12120.

13. If a ship sails east from port 350 miles a day for 4 days, and is then driven west 289 miles, how far is she from port? *Ans.* 1111 miles.

14. John has 31 marbles; James has 6 times as many; Henry has as many as John and James together. How many marbles have they all? *Ans.* 434 marbles.

15. What cost 100 wagons, at $75 apiece?

16. Find the product of 8089 and 9007. Subtract this product increased by 377, from 100000000. *Ans.* 27142000.

17. Trinity Church, N.Y., is 284 feet high. St. Peter's at Rome is 166 feet higher; what is its height?

18. A man who has $1201 in the bank, draws out enough to pay for 5 lots of land at $195 a lot. How much is left in the bank? *Ans.* $226.

19. North America contains 8000000 square miles; South America, 7000000; Asia, 1000000 more than both Americas together. How many square miles in Asia? *Ans.* 16000000.

20. An army of 12100 men lost 631 in killed and wounded, and twice that number taken prisoners. How many were left? *Ans.* 10207 men.

21. In a school of 127 boys, four times as many study Arithmetic as study Latin. The Latin class numbers 28; how many study Arithmetic? *Ans.* 112 boys.

22. What is the cost of a carriage and span of horses, if each horse cost $235, and the carriage $679? *Ans.* $1149.

23. A boat makes 208 trips in a season, and carries on an average 106 passengers each trip; how many does she carry in all? *Ans.* 22048 passengers.

24. Find the cost of 53 mules, at $100 apiece.

25. A grocer who has 15 firkins of butter, containing 56 pounds each, sells four of his customers 240 pounds. How much has he left? *Ans.* 600 pounds.

5

DIVISION.

83. If 2 apples can be bought for 1 cent, how many cents will 6 apples cost?

2 apples cost 1 cent; hence 6 apples will cost as many cents as 2 is contained times in 6. Here we are required to find how many times 2 is contained in 6. This process is called Division.

84. Division is the process of finding how many times one number is contained in another.

Division Table.

Any number is contained in 0, 0 times.

1 in 1, once.	2 in 2, once.	3 in 3, once.
1 in 2, twice.	2 in 4, twice.	3 in 6, twice.
1 in 3, 3 times.	2 in 6, 3 times.	3 in 9, 3 times.
1 in 4, 4 times.	2 in 8, 4 times.	3 in 12, 4 times.
1 in 5, 5 times.	2 in 10, 5 times.	3 in 15, 5 times.
1 in 6, 6 times.	2 in 12, 6 times.	3 in 18, 6 times.
1 in 7, 7 times.	2 in 14, 7 times.	3 in 21, 7 times.
1 in 8, 8 times.	2 in 16, 8 times.	3 in 24, 8 times.
1 in 9, 9 times.	2 in 18, 9 times.	3 in 27, 9 times.
4 in 4, once.	5 in 5, once.	6 in 6, once.
4 in 8, twice.	5 in 10, twice.	6 in 12, twice.
4 in 12, 3 times.	5 in 15, 3 times.	6 in 18, 3 times.
4 in 16, 4 times.	5 in 20, 4 times.	6 in 24, 4 times.
4 in 20, 5 times.	5 in 25, 5 times.	6 in 30, 5 times.
4 in 24, 6 times.	5 in 30, 6 times.	6 in 36, 6 times.
4 in 28, 7 times.	5 in 35, 7 times.	6 in 42, 7 times.
4 in 32, 8 times.	5 in 40, 8 times.	6 in 48, 8 times.
4 in 36, 9 times.	5 in 45, 9 times.	6 in 54, 9 times.

83. Repeat the example. What are we here required to do? What is this process called?—84. What is Division? How many times is any number contained in 0? Recite the Table.

7 in 7, once.	8 in 8, once.	9 in 9, once.
7 in 14, twice.	8 in 16, twice.	9 in 18, twice.
7 in 21, 3 times.	8 in 24, 3 times.	9 in 27, 3 times.
7 in 28, 4 times.	8 in 32, 4 times.	9 in 36, 4 times.
7 in 35, 5 times.	8 in 40, 5 times.	9 in 45, 5 times.
7 in 42, 6 times.	8 in 48, 6 times.	9 in 54, 6 times.
7 in 49, 7 times.	8 in 56, 7 times.	9 in 63, 7 times.
7 in 56, 8 times.	8 in 64, 8 times.	9 in 72, 8 times.
7 in 63, 9 times.	8 in 72, 9 times.	9 in 81, 9 times.

10 in 10, once.	11 in 11, once.	12 in 12, once.
10 in 20, twice.	11 in 22, twice.	12 in 24, twice.
10 in 30, 3 times.	11 in 33, 3 times.	12 in 36, 3 times.
10 in 40, 4 times.	11 in 44, 4 times.	12 in 48, 4 times.
10 in 50, 5 times.	11 in 55, 5 times.	12 in 60, 5 times.
10 in 60, 6 times.	11 in 66, 6 times.	12 in 72, 6 times.
10 in 70, 7 times.	11 in 77, 7 times.	12 in 84, 7 times.
10 in 80, 8 times.	11 in 88, 8 times.	12 in 96, 8 times.
10 in 90, 9 times.	11 in 99, 9 times.	12 in 108, 9 times.

85. The number to be divided, is called the **Dividend**; that by which we are to divide, the **Divisor.**

The result, or number obtained by dividing, is called the **Quotient.** It shows how many times the divisor is contained in the dividend.

2 is contained in 6, 3 times; 2 is the divisor, 6 the dividend, 3 the quotient.

86. The divisor is not always contained an exact number of times in the dividend. It may go a certain number of times, and some over. What is left over is called the **Remainder.**

5 is contained in 15 exactly 3 times. In 16 it goes 3 times, and 1 over; 3 is the quotient, 1 the remainder.

85. What is the number to be divided called? What is that by which we are to divide called? What is the result obtained by dividing called? What does the quotient show? Illustrate these definitions.—86. What is meant by the Remainder? Give an example.

87. Division is denoted by a short horizontal line between two dots ÷. This sign indicates that the number before it is to be divided by the one after it. 6÷2 is read, and denotes, *six divided by two.*

MENTAL EXERCISES.

How many times is 3 contained in 9? 10 in 40? 7 in 56? 11 in 99? 4 in 32? 9 in 81? 6 in 42? 12 in 36? 2 in 18? 11 in 55? 5 in 30? 12 in 96? 8 in 72? 7 in 0?

How many times is 4 contained in 23? (*Ans.* 5 *times, and* 3 *over.*) 2 in 5? 6 in 27? 9 in 20? 8 in 39? 3 in 10? 4 in 29? 5 in 38? 10 in 44? 6 in 59? 12 in 75?

What is the quotient, and what the remainder, in the following? 34÷4. 31÷6. 22÷8. 108÷12. 37÷5. 13÷3. 65÷9. 0÷4. 89÷10. 8÷5. 43÷8. 72÷9. 47÷11. 72÷8. 24÷3. 24÷8. 64÷12. 28÷10. 81÷11.

1. If 64 cherries are divided into 8 equal piles, how many will there be in each pile?

MODEL.—As many cherries as 8 is contained times in 64, or 8. Answer, 8 cherries.

2. Twenty-one cents are distributed equally among 7 beggars. What is the share of each?

3. How many coats, at $5 apiece, can I buy for $20?

4. John's father gave him 18 pears, to be divided equally among his two sisters and himself. How many did each get?

5. How many bushels are there in 16 pecks, there being 4 pecks in 1 bushel?

6. Twelve make a dozen. How many dozen in 24?

7. Allowing 8 yards of calico for a dress, how many dresses can be made out of 40 yards?

---

87. How is division denoted? What does this sign indicate?

8. If a dozen pineapples are sold for 96 cents, how much is that apiece?

9. How many stage-coaches, carrying 9 persons each, will be needed to carry 45 passengers?

10. How many ten-ounce bottles will it take to hold 40 ounces of alcohol?

11. There are 49 days in 7 weeks. How many days are there in 1 week?

12. If a family use 42 quarts of milk in a week, how much do they use in a day?

13. If 11 sheep yield 33 pounds of wool, what is the average yield for each sheep?

14. How many caps, at $2 each, can you buy for $12?

15. If I go 50 miles in 10 hours, what is my rate per hour?

16. How many pair will 16 chickens make?

17. If 48 trees are planted in 6 equal rows, how many trees will there be in a row?

18. Eighty-four eggs make how many dozen?

19. Twelve yards come in a piece of ribbon. How many pieces will 60 yards of ribbon make?

EXAMPLES FOR THE SLATE.

88. Divide 27036 by 3.

Here the divisor is a single figure. Set it at the left of the dividend, with a curved line between.

3) 27036

In dividing, always begin at the left. See how many times the divisor is contained in each figure of the dividend, and set the quotient under the figure divided.

---

88. When the divisor is a single figure, where is it set? Where must we always begin in dividing? How do we find the quotient?

3 is not contained in 2. See, then, how often it will go into 27, the first two figures. 3 in 27, 9 times. Set down 9 under 7, the right-hand figure of the two divided.

```
3)27036
  9012
```

3 in 0, 0 times. Set it down, and remember that 0 must never be omitted in the quotient, unless it is the first figure.

3 in 3, once; set down 1. 3 in 6, twice; set down 2. Answer, 9012.

89. In the following examples, divide as above:—

(1)	(2)	(3)	(4)
2)148264	3)219063	4)28084	5)40500

5. Divide thirty-six hundred and six, by six.

6. Divide fifty-six million by 7: by 8.

Carrying.

90. When, after having divided all the figures of the dividend, there is a remainder, set it down as such. But if before this a remainder occurs, prefix it (in the mind) to the next figure of the dividend, and continue the division.

This prefixing is called **Carrying.**

Example.—Divide 265231 by 6.

```
   2 1 3
6)265231
  44205 and 1 remainder.
```

6 is not contained in 2. 6 in 26, 4 times and 2 over; set down 4 under the 6, and carry 2. 6 in 25, 4 times and 1 over; set down 4 and carry 1. 6 in 12, twice; set down 2. 6 in 3, 0 times and 3 over; set down 0 and carry 3. 6 in 31, 5 times and 1 over; set down 5 in the quotient, and 1 as remainder. Answer, 44205 and 1 remainder.

Go through these steps in the given example.—90. What is meant by Carrying in division? When is the remainder set down? Show how we carry, in the given example.

91. Rule for Carrying.—*If, in dividing any figure of the dividend except the last, a remainder occurs, prefix it (in the mind) to the next figure to be divided, and continue the division.*

92. When the divisor is greater than the figure to be divided, set 0 in the quotient, and carry the latter figure.

Proof of Division.

93. Multiply the quotient by the divisor, and add in the remainder, if there is one. If the result equals the dividend, the work is right.

Example.—Prove the example given on the last page.

Multiply the quotient 44205, by the divisor 6. Add in the remainder, 1. The result is 265231, which is the same as the dividend. Hence the work is right.

```
 44205
     6
------
265230
     1
------
265231
```

EXAMPLES FOR THE SLATE.

Find the quotient and remainder. Prove each example.

1. 50736÷2. *Ans.* 25368.
2. 271216÷8. *Ans.* 33902.
3. 124555806÷3.
4. 2501024085÷5.
5. 5068595742÷7.
6. 117360194÷9. *Rem.* 5.
7. 219007637÷4. *Rem.* 1.
8. 470169628÷8. *Rem.* 4.
9. 401013116÷6. *Rem.* 2.
10. 303913211÷5. *Rem.* 1.
11. Divide one million four hundred thousand six hundred and twelve, by 9. *Ans.* 155623, 5 rem.
12. The dividend is fifteen million; the divisor is seven; find the quotient. *Rem.* 1.

---

91. Give the rule for carrying in division.—92. What must be done when the divisor is greater than the figure to be divided?—93. How is division proved? Prove the example just given.

13. How often is 3 times 2 contained in one hundred and forty million and four? *Ans.* 23333334 times.

14. If it costs $17068 to build four miles of plank road, what will one mile cost?

15. A man leaves $317520 to be divided equally between his two children. What is the share of each?

16. John gathers 761 nuts, and Jacob 843. If they share them equally, how many will each have? *Ans.* 802 nuts.

17. How many six-acre fields can be laid out in a plantation of 1488 acres?

94. The mode of dividing shown above is called **Short Division.** In Short Division, the carrying is done in the mind, and the quotient is written under the dividend.

Dividing by two or more figures.

95. Divide by 10, 11, and 12, by short division.

Example.—Divide 1084608 by 12.

12 is not contained in 1, or in 10. Hence we take the first three figures. 12 in 108, 9 times; set down 9 under 8, the right-hand figure of those divided.

```
12)1084608
    90384
```

12 in 4, 0 times; set down 0. 12 in 46, 3 times and 10 over; set down 3 and carry 10. 12 in 100, 8 times and 4 over; set down 8 and carry 4. 12 in 48, 4 times; set down 4. Answer, 90384.

Find the value of the following:—

1. 8314607÷10. *Rem.* 7.
2. 9923166÷11. *Rem.* 0.
3. 1073285÷10. *Rem.* 5.
4. 1109964÷12. *Ans.* 92497.
5. 1018193÷11. *Ans.* 92563.
6. 1198872÷12. *Ans.* 99906.

---

94. What is this mode of dividing called? In Short Division, how is the carrying done? Where is the quotient written?—95. How are we to divide by 10, 11, 12? Give an example.

96. When the divisor is over 12, proceed by what is called **Long Division.**

97. In Short Division, we set the quotient *under* the dividend. In Long Division, we place it *at the right*, with a curved line between.

In Short Division, we find what is to be carried, and prefix it to the next figure, *in the mind.* In Long Division, we multiply, subtract, and prefix as before; but we *write down* all the figures used.

EXAMPLE.—Divide 9594 by 39.

```
       39)9594(246
39×2=  78
       ---
       179
39×4=  156
       ---
        234
39×6=   234
        ---
```

The quotient is now to be set at the right of the dividend. Beginning at the left of the dividend, take as many figures as are required to contain the divisor—in this case, two. We find on trial that 39 is contained in 95 twice. Set 2 in the quotient as the first figure.

Multiply the divisor by this 2. Twice 39 is 78; set the product under 95, and subtract. The remainder is 17, which (as in short division) we prefix, by bringing down 9, the next figure of the dividend.

Now repeat the same steps. Find how often 39 is contained in 179. It goes 4 times. Set 4 in the quotient; multiply the divisor by it; set the product under 179, and subtract. The remainder is 23, to which bring down 4, the next figure of the dividend.

Find how often 39 is contained in 234. It goes 6 times. Set 6 in the quotient; multiply the divisor by it; set the product under 234, and subtract. There is no remainder. As we have now brought down all the figures of the dividend, the sum is finished. Answer, 246.

95, 179, and 234, are called **Partial Dividends.**

---

96. What process do we use when the divisor is over 12?—97. Show how Long Division differs from Short Division. Divide 9594 by 39, explaining the several steps.

98. We may not always, on the first trial, get the right quotient figure.

If, when we multiply the divisor by any quotient figure, the product comes greater than the partial dividend, the quotient figure is too great, and must be diminished.

If, on the other hand, we have a remainder greater than the divisor, the quotient figure is too small, and must be increased.

```
39)9594(3
   117
   ———
```

EXAMPLES.—In the last example, if we say 39 is contained 3 times in 95, we get a product greater than the partial dividend, and can not subtract. We must therefore diminish the quotient figure.

If we say it is contained once, on multiplying and subtracting we get a remainder greater than the divisor. We must therefore increase the quotient figure.

```
39)9594(1
   39
   ——
   56
```

99. For every figure of the dividend brought down, a figure must be placed in the quotient. To prevent mistakes, it is well to place a dot under each figure as it is brought down.

100. When the divisor is not contained in the partial dividend, set 0 in the quotient, and bring down the next figure of the dividend.

If several figures are brought down before the divisor will go into the partial dividend, set a 0 in the quotient for each.

What are the numbers formed each time after a figure is brought down called?—98. In what are we liable to make mistakes? When may we know that the quotient figure is too great? When, that it is too small? Give examples.—99. To prevent mistakes in bringing down the figures, what is recommended?—100. When the divisor is not contained in the partial dividend, what must be done?

EXAMPLE.—Divide 172602 by 86.

```
86)172602(2007
   172
   ---
      602
      602
      ---
```

Here our first remainder is 0. Bring down 6. 86 in 6, 0 times; set 0 in the quotient, and bring down the next figure. 86 in 60, 0 times; set another 0 in the quotient, and bring down 2, the next figure. 86 in 602, 7 times. Set 7 in the quotient; multiply and subtract. Answer, 2007.

EXAMPLES FOR THE SLATE.

Prove each example. In the first nine, if your work is right, the remainder will come equal to the quotient.

1. Divide 6765 by 122.
2. Divide 80157 by 346.
3. Divide 34075 by 234.
4. Divide 159750 by 425.
5. Divide 269459 by 576.
6. Divide 39298 by 801.
7. Divide 466281 by 926.
8. Divide 854080 by 1087.
9. Divide 3054436 by 2953.

10. 259828÷34. *Ans.* 7642.
11. 70338÷57. *Ans.* 1234.
12. 539902÷23. *Rem.* 0.
13. 470112÷354. *Ans.* 1328.
14. 16199569÷1848. *Rem.* 1.
15. 14003794÷6974. *Rem.* 2.
16. 13373585÷2643. *Rem.* 5.
17. 1238025÷1536. *Rem.* 9.
18. 789783÷19263. *Rem.* 0.

19. Divide 24280295 by 9684. *Rem.* 2507.
20. Divide 1521808704 by 234. *Ans.* 6503456.
21. Divide 5763447 by 678509. *Rem.* 335375.
22. Divide 24280830966 by 604. *Rem.* 162.
23. Divide 48423768038 by 807009. *Rem.* 2.
24. Divide 22687013 by 310781. *Ans.* 73.
25. Divide 78204820 by 8679548. *Rem.* 88888.
26. Divide 36132610 by 8603. *Rem.* 10.
27. Divide 1651877824 by 6503456. *Ans.* 254.
28. Divide 7807085 by 945. *Rem.* 440.
29. Divide 4848708 by 808. *Rem.* 708.
30. Divide 1917183 by 9682. *Rem.* 147.

General Rule for Division.

1. *Set the divisor at the left of the dividend, with a line between.*

2. *Take as many figures at the left of the dividend as will contain the divisor, and find how many times it will go into them.*

3. *If the divisor is* 12 *or less, set this first quotient figure under the figure divided, or under the right-hand figure of those divided, if more than one are taken. Divide each figure of the dividend in turn, carrying what is over, and setting each quotient figure under the figure divided.*

4. *If the divisor is over* 12, *set the first quotient figure at the right of the dividend. Multiply the divisor by it, and subtract the product from the partial dividend.*

5. *Bring down the next figure of the dividend. Find the next quotient figure, multiply, and subtract, as before. Go on thus, till all the figures of the dividend are brought down.*

6. *If any partial dividend is too small to contain the divisor, set* 0 *in the quotient, bring down the next figure, and go on as before.*

EXAMPLES FOR THE SLATE.

1. The greatest height reached by man in a balloon is 23027 feet. How many miles is this, allowing 5280 feet to the mile? *Ans.* 4 miles, 1907 feet.

2. The earth's circumference is about 25000 miles. How many days would it take a person to traverse it, going at the rate of 125 miles a day?

3. How many barrels of apples, at $3 a barrel, can be bought for $2568? If one tree produces 8 barrels, how many trees will it take to yield this quantity? *Ans.* 107 trees.

4. If a ship bound from New York to Canton sails 9225 miles, and is 75 days making the voyage, how many miles a day does she average?

5. The earth's distance from the sun is about 95000000 miles. It takes a sunbeam about 8 minutes to reach the earth; how many miles does light travel in a minute? *Ans.* 11875000 miles.

6. A flour barrel holds 196 pounds of flour. How many barrels will it take to hold 406700 pounds? *Ans.* 2075 bbl.

7. If 47 miles of railroad cost $1815375, what is the cost per mile? *Ans.* $38625.

8. The product of two factors is 392328687. One of the factors is 46391; what is the other?* *Ans.* 8457.

9. Twenty-six times a certain number is 1580540. What is the number? *Ans.* 60790.

10. In 1850 there were 2526 newspapers and periodicals in the United States, of which 426409978 copies were printed yearly. What was their average yearly circulation? *Ans.* Over 168808 copies.

11. If 218669 bushels of rye are stored in equal lots in 11 warehouses, how much is there in each? *Ans.* 19879 bu.

12. How long will 15000 pounds of flour last a garrison of 250 men, allowing them 750 pounds a day? *Ans.* 20 days.

13. How many impressions can a printing press make in one hour, if it makes 633240 impressions in one day of 24 hours? *Ans.* 26385 impressions.

---

* NOTE.—When a product and one of its factors are given, to find the other, divide the product by the given factor.

$3 \times 9 = 27$ Then $27 \div 3 = 9$ Or, $27 \div 9 = 3$

Naughts at the right of the divisor.

101. *When there are naughts at the right of the divisor, cut them off, and also as many figures at the right of the dividend.*

Divide the remaining figures of the dividend by those of the divisor. If there is a remainder, annex to it the figures cut off from the dividend for the true remainder; if not, the figures cut off are the true remainder.

Divisor Dividend Quo.	Divisor Dividend Quo.	Divisor Dividend
26Ø)539Ø(20	47ØØ)2371Ø(5	3ØØØ)189ØØ1
52	235	63 Quo.
19	2	
Ans. 20, 190 rem.	*Ans.* 5, 216 rem.	*Ans.* 63, 1 rem.

102. Dividing a number by 1 does not change its value. Hence, *to divide a number by* 10, 100, 1000, *&c., simply cut off as many figures at the right of the dividend as there are naughts in the divisor. The remaining figures are the quotient; those cut off, the remainder.*

3500÷10=350 3500÷100=35 3500÷1000=3, 500 rem.

EXAMPLES FOR THE SLATE.

1. Divide 186740 by 10.
2. Divide 729500 by 100.
3. Divide 729500 by 1000.
4. Divide 729500 by 10000.
5. Divide 901109 by 1000.
6. 2294003÷3700. *Rem.* 3.
7. 854009÷14000. *Rem.* 9.
8. 1782000÷9900. *Rem.* 0.
9. 405349÷490. *Rem.* 119.
10. 253579÷510. *Rem.* 109.

101. How must we proceed when there are naughts at the right of the divisor?—102. What is the effect of dividing a number by 1? Give the rule for dividing by 10, 100, 1000, &c.

11. About three million tons of iron are produced yearly in Great Britain, which is twenty times as much as is produced annually in Sweden. How much is produced in Sweden? *Ans.* 150000 tons.

12. A ship bound from Boston to Australia, a voyage of 13000 miles, sails at the rate of 100 miles a day. How long is she on the passage?

13. The product of two factors is 238700. One of the factors is 770; what is the other? *Ans.* 310.

14. A cotton crop of 26320 pounds was put up in bales averaging 560 pounds each. How many bales did it make?

15. Sound moves 1120 feet in a second. How long before we hear a cannon, fired at a distance of 12320 feet?

MISCELLANEOUS EXAMPLES FOR THE SLATE.

To find a sum, add.
To find a difference, subtract.
To find a product, multiply.
To find a quotient, divide.

1. Find the sum, then the difference, then the product, then the quotient, of 125 and 875.

2. A person who had 200 acres in one state, 1173 in another, and 127 in a third, divided his land equally among his wife and four sons. How much did each receive?

Find how much land he had in all. Among how many did he divide it?

3. If a clock ticks 3600 times in 1 hour, how often will it tick in two days of 24 hours each? *Ans.* 172800 times.

4. How many times is 20×50 contained in 50×20?

5. A barrel of flour contains 196 pounds. If a family of 5 persons use 4 pounds of flour a day, how long will a barrel last them? *Ans.* 49 days.

6. Three men put in $650 each, and invest the whole in salt, at $2 a sack. How many sacks do they buy? *Ans.* 975.

7. A merchant worth $43000, after losing $1750 in trade, invested the rest in land, at $10 an acre. How many acres did he buy? *Ans.* 4125 acres.

8. What is the difference between 16 × 49 and 49 × 8 × 2?

9. A lady who had eight children, lost two of them. Among the survivors she divided her property; which consisted of $1200 cash, $21550 in stock, and $7250 in bonds. What was the share of each? *Ans.* $5000.

10. The sum of three numbers is 27846. One of the numbers is 11587; the second is 596; what is the third?

Find the sum of the two given numbers. The third will be the difference between their sum and the whole sum.

Ans. 15663.

11. A man leaves his three sons $36000. To the first he leaves $14655; to the second, $9875; how much does he leave the third? *Ans.* $11470.

12. A soapboiler makes 5977 pounds of soap, of which 657 pounds are soft soap, while the rest is in bars. He wishes to pack his bar soap in boxes holding 70 pounds each. How many boxes must he procure? *Ans.* 76 boxes.

13. How much must I add to 40 times 60, in order to make 3000?

How much is 40 × 60? What is the difference between this and 3000?

14. A farmer wishes to buy some land for $720. He lays up $9 a week for one year, or 52 weeks. How much does he still need? *Ans.* $252.

15. A grocer, having 3 boxes of tea holding 80 pounds each, repacks it in six-pound boxes. How many boxes will it fill? *Ans.* 40 boxes.

16. Two men go from the same point in opposite directions, one 18 miles a day, the other 23. When they are 369 miles apart, how many days have they travelled?

How far apart are they at the end of one day? How many times is this number contained in 369?

17. A company building a railroad, pay $377235 for labor, and $147690 for other expenses. The road is 15 miles long; what is the cost per mile? *Ans.* $34995.

18. A farmer has 917 sheep in one field; 189 in another; 276 in a third; and 379 in a fourth. If he divides them into three equal lots, how many will there be in each? *Ans.* 587.

19. The Vice-President's salary is $8000 a year. If he spends $15 a day, how much will he lay up during his four years' term, allowing 365 days to the year? *Ans.* $10100.

20. Two men go from the same point in the same direction, one at the rate of 41 miles a day, and the other 24. When they are 493 miles apart, how many days have they travelled? *Ans.* 29 days.

21. I buy 29 horses for $1885, and sell them at a profit of $290. How much do I get apiece? *Ans.* $75.

22. A planter raises 25963 pounds of cotton, and buys up 35409 pounds more. If he packs the whole in bales of 458 pounds, how many bales will he have? *Ans.* 134 bales.

23. Roger Bacon, the inventor of spectacles, lived to the age of 78, and died 350 years before Sir Isaac Newton was born. If Newton was born in 1642, what was the year of Roger Bacon's birth? *Ans.* 1214.

24. How many hogsheads holding 1295 pounds each will it take to contain 76405 pounds of sugar? *Ans.* 59 hogsheads.

25. Five partners make $1500 by a speculation. One of them divides his share between his two daughters. What does each daughter get? *Ans.* $150.

26. Find the value of 637+8465367−487. *Ans.* 8465517.

27. How much is 986743+9767+98437? *Ans.* 1094947.

28. Find the value of 64372159−8524384. *Ans.* 55847775.

29. Find the product of 79, 86, and 54. *Ans.* 366876.

30. Divide 1824962163 by 26837548. *Ans.* 68, 8899 rem.

FRACTIONS.

103. When a whole is divided into two equal parts, each of these parts is called one Half.

When a whole is divided into three equal parts, one of these parts is called one Third; two are two Thirds; &c.

When a whole is divided into four equal parts, one of these parts is called one Fourth (or Quarter); two are called two Fourths; three, three Fourths; &c.

In the same way we get Fifths, Sixths, Sevenths, &c., by dividing a whole into *five*, *six*, *seven*, &c., equal parts. The name is taken from the number of equal parts into which the whole is divided.

104. The value of the part varies according to the number of parts into which the whole is divided. The more parts it is divided into, the smaller they must be.

Half		Half	
Third	Third	Third	
Fourth	Fourth	Fourth	Fourth

One half of a thing is greater than one third; one third is greater than one fourth.

103. When a whole is divided into equal parts, if there are two, what is each called? If there are three, what is each called? What are two called? If there are four equal parts, what is each called? What are three such parts called? How do we get Fifths, Sixths, Sevenths, &c.? From what is the name taken?—104. According to what does the value of the part vary? Which is greater, one half or one third? One fourth or one third?

105. These equal parts into which a whole is divided, are called **Fractions.**

Writing of Fractions.

106. Learn how fractions are written.

One half	$\frac{1}{2}$	Five thirteenths	$\frac{5}{13}$
One third	$\frac{1}{3}$	Three twenty-seconds	$\frac{3}{22}$
One fourth (quarter)	$\frac{1}{4}$	Twenty sixty-firsts	$\frac{20}{61}$
One tenth	$\frac{1}{10}$	Nine three-hundredths	$\frac{9}{300}$
One two-hundredth	$\frac{1}{200}$	Three thousandths	$\frac{3}{1000}$
One thousandth	$\frac{1}{1000}$	Six twelve-hundredths	$\frac{6}{1200}$

107. A fraction, therefore, when written, consists of two numbers, one below the other, with a line between.

The number below the line is called the **Denominator.** It shows into how many equal parts the whole is divided, and therefore gives name to these parts.

The number above the line is called the **Numerator.** It shows how many of the equal parts denoted by the Denominator are taken.

The Numerator and the Denominator, taken together, are called the **Terms** of the fraction.

$\frac{5}{6}$ is a fraction. 5 and 6 are its Terms. 6 is the Denominator, and shows that the whole is divided into *six* equal

---

105. What are such equal parts of a whole called?—106. Write one half; one third; one quarter; &c.—107. Of what does a written fraction consist? What is the number below the line called? What does the Denominator show? What is the number above the line called? What does the Numerator show? What are Numerator and Denominator, taken together, called? Name the terms of the fraction *five sixths*. Name its numerator; its denominator.

parts, making each part one *sixth*. 5 is the Numerator, and shows that *five* of these equal parts are taken.

In reading, name the numerator first,—*five sixths*. Always pronounce *th* distinctly, in naming the part denoted by the denominator,—*six*TH.

EXERCISE.

Read these fractions. Then name the numerator and the denominator, and tell what each shows.

$\frac{4}{9}$; $\frac{7}{12}$; $\frac{80}{99}$; $\frac{5}{116}$; $\frac{27}{301}$; $\frac{19}{2002}$; $\frac{501}{3010}$; $\frac{6019}{10000}$.

Write the following fractions in figures:—

1. Seven sixteenths.
2. Nine tenths.
3. Two five-hundredths.
4. Eleven billionths.
5. Eighty hundredths.
6. Three millionths.
7. One thousandth.
8. Five seventy-seconds.
9. Twenty-one ninetieths.
10. Sixty ten-thousandths.
11. Sixteen twelve-hundred-and-ninety-firsts.
12. Five hundred four-hundred-thousandths.

Definitions.

108. A **Fraction** is one or more of the equal parts into which a whole is divided; as, $\frac{1}{3}$, $\frac{2}{3}$.

109. A **Proper Fraction** is one whose numerator is less than its denominator; as, $\frac{2}{3}$, $\frac{19}{20}$.

110. An **Improper Fraction** is one whose numerator is equal to or greater than its denominator; as, $\frac{3}{3}$, $\frac{24}{20}$.

111. A **Mixed Number** is one that consists of a whole number and a fraction; as, $7\frac{1}{2}$ (*seven and a half*).

In reading, which of the terms must be named first?—108. What is a Fraction?—109. What is a Proper Fraction?—110. What is an Improper F.action?—111. What is a Mixed Number?

Fractional Parts of Whole Numbers.

112. A fraction indicates division. The fractional line is the line used in the sign of division ÷. The dividend is written in place of the dot above the line; the divisor, in place of the dot below it. Hence,

To find $\frac{1}{2}$, divide by 2.	To find $\frac{1}{4}$, divide by 4.
To find $\frac{1}{3}$, divide by 3.	To find $\frac{1}{5}$, divide by 5.

And generally, *To find one of the equal parts denoted by a denominator, divide by the denominator.*

MENTAL EXERCISES.

1. How much is $\frac{1}{2}$ of 14? Of 8? Of 18? Of 2?
2. How much is $\frac{1}{5}$ of 20? Of 45? Of 10? Of 50?
3. Find $\frac{1}{7}$ of 21. Of 56. Of 63. Of 14. Of 35.
4. Find $\frac{1}{9}$ of 81. Of 18. Of 63. Of 27. Of 54.
5. $\frac{1}{10}$ of 80? Of 20? Of 100? Of 70? Of 30?
6. $\frac{1}{12}$ of 72? Of 48? Of 24? Of 12? Of 84?
7. $\frac{1}{8}$ of 16? Of 80? Of 48? Of 24? Of 72?
8. $\frac{1}{11}$ of 22? Of 66? Of 99? Of 33? Of 55?
9. $\frac{1}{3}$ of 15? $\frac{1}{6}$ of 54? $\frac{1}{4}$ of 28? $\frac{1}{12}$ of 36? $\frac{1}{6}$ of 36?
10. $\frac{1}{4}$ of 24? $\frac{1}{6}$ of 48? $\frac{1}{3}$ of 27? $\frac{1}{3}$ of 18? $\frac{1}{11}$ of 11?
11. $\frac{1}{6}$ of 30? $\frac{1}{3}$ of 9? $\frac{1}{4}$ of 16? $\frac{1}{6}$ of 24? $\frac{1}{6}$ of 12?
12. $\frac{1}{10}$ of 50? $\frac{1}{100}$ of 500? $\frac{1}{1000}$ of 4000? $\frac{1}{10}$ of 640?
13. $\frac{1}{7}$ of 70? $\frac{1}{2}$ of 6? $\frac{1}{8}$ of 32? $\frac{1}{5}$ of 15? $\frac{1}{3}$ of 3?
14. $\frac{1}{6}$ of 42? $\frac{1}{20}$ of 20? $\frac{1}{11}$ of 77? $\frac{1}{8}$ of 56? $\frac{1}{12}$ of 96?

---

112. What does a fraction indicate? With what does the fractional line correspond? Where is the dividend written, and where the divisor? How can we find one half? One third? Give the rule for finding one of the equal parts denoted by a denominator

113. *To find more than one of the equal parts into which a whole is divided, first find one in the way just shown; then multiply this by the number of parts required.*

EXAMPLES.—1. How much is $\frac{3}{5}$ of 40?

One fifth of 40 is 8; and *three* fifths are 3 times 8, or 24. Answer, 24.

2. A boy having 27 cents earned $\frac{4}{9}$ as much more. How many cents did he earn?

He earned $\frac{4}{9}$ of 27 cents. *One* ninth of 27 cents is 3 cents; and *four* ninths are 4 times 3 cents, or 12 cents. Answer, 12 cents.

MENTAL EXEROISES.

1. How much is $\frac{2}{3}$ of 18? $\frac{3}{7}$ of 21? $\frac{1}{12}$ of 108? $\frac{5}{12}$ of 108? $\frac{7}{12}$ of 108? $\frac{5}{6}$ of 42? $\frac{3}{4}$ of 4? $\frac{5}{9}$ of 36? $\frac{2}{5}$ of 20?

2. How much is $\frac{5}{11}$ of 22? $\frac{4}{11}$ of 44? $\frac{5}{9}$ of 72? $\frac{2}{5}$ of 40? $\frac{3}{8}$ of 48? $\frac{7}{10}$ of 30? $\frac{2}{11}$ of 66? $\frac{9}{10}$ of 50? $\frac{5}{8}$ of 64?

3. How much is $\frac{3}{100}$ of 200? $\frac{3}{9}$ of 27? $\frac{5}{4}$ of 24? $\frac{8}{5}$ of 15? $\frac{7}{6}$ of 18? $\frac{12}{9}$ of 81? $\frac{2}{12}$ of 60? $\frac{7}{11}$ of 55? $\frac{6}{12}$ of 72?

4. A man having 48 cows sold $\frac{3}{4}$ of them. How many did he sell?

5. At 81 cents a pound, what will $\frac{8}{9}$ of a pound of tea cost?

6. A boy having 20 cents gave away one tenth of them and spent two tenths. How many cents had he left? *Ans.* 14c.

EXAMPLES FOR THE SLATE.

1. Find $\frac{3}{14}$ of 280. *Ans.* 60.
2. Find $\frac{11}{49}$ of 441. *Ans.* 99.
3. Find $\frac{27}{35}$ of 1610. *Ans.* 1242.
4. Find $\frac{43}{60}$ of 4200. *Ans.* 3010.
5. Find $\frac{6}{145}$ of 8410. *Ans.* 348.
6. Find $\frac{1}{69}$ of 8832. *Ans.* 128.
7. Find $\frac{11}{296}$ of 7400. *Ans.* 275.
8. Find $\frac{7}{10000}$ of 6100000.

---

113. How can we find more than one of the equal parts into which a whole is divided?

114. A fraction indicates division. Hence, in division, when there is a remainder, in stead of writing it as such beside the quotient, it is usual to place it over the divisor in the form of a fraction.

EXAMPLE.—Find $\frac{1}{7}$ of 45. Answer, $6\frac{3}{7}$.

$$7\overline{)45}$$
6, and 3 rem.

115. Reducing whole numbers to halves, &c.

2 halves make 1 whole.	5 fifths make 1 whole.
3 thirds make 1 whole.	6 sixths make 1 whole.
4 fourths make 1 whole.	7 sevenths make 1 whole.

Hence, *To reduce to halves, multiply by* 2.
To reduce to thirds, multiply by 3.
To reduce to fourths, multiply by 4.
To reduce to fifths, multiply by 5, &c.

EXAMPLES FOR THE SLATE.

1. How many thirds in 2901? *Ans.* 8703 thirds; or, $\frac{8703}{3}$.
2. If I cut 45 pies into halves, how many halves have I?
3. How many quarters of beef will 96 oxen make?
4. How many tenths of an inch in 72 inches?
5. Reduce 497 to fiftieths. Reduce 876 to twelfths.
6. A cent is one hundredth of a dollar. How many cents are there in $75? *Ans.* 7500 cents.
7. A mill is $\frac{1}{1000}$ of a dollar. How many mills in $34?
8. Find one eleventh of 25304618. *Ans.* $2300419\frac{9}{11}$.
9. Find one thousandth of 19001. *Ans.* $19\frac{1}{1000}$.
10. Find one sixty-eighth of 5375. *Ans.* $79\frac{3}{68}$.

---

114. What does a fraction indicate? Hence, in division, how is it usual to write a remainder?—115. How do we reduce a whole number to halves? To thirds? To fifths? To twentieths?

Reduction of Fractions.

116. Reducing a fraction is changing its form without changing its value.

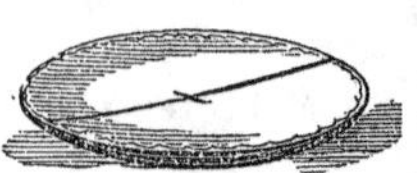

If we divide a pie into 2 equal parts, each part is called $\frac{1}{2}$. If we divide each half into 2 equal parts, we get four parts in all, and each is therefore $\frac{1}{4}$ of the whole.

Now two of these fourths are made from one half, and are therefore equal to one half. $\frac{2}{4}$, therefore, may be written $\frac{1}{2}$,—or, as it is generally expressed, *reduced to* $\frac{1}{2}$.

Reduction to Lowest Terms.

117. We have just seen that $\frac{2}{4} = \frac{1}{2}$. Now what operation performed on $\frac{2}{4}$ gives $\frac{1}{2}$? Dividing both numerator and denominator by 2.

$$\frac{2 \div 2}{4 \div 2} = \frac{1}{2}$$

We have just seen that $\frac{1}{2} = \frac{2}{4}$. Now what operation performed on $\frac{1}{2}$ gives $\frac{2}{4}$? Multiplying both numerator and denominator by 2.

$$\frac{1 \times 2}{2 \times 2} = \frac{2}{4}$$

The value of a fraction, therefore, is not changed by dividing or multiplying both numerator and denominator by the same number.

118. A fraction is said to be *in its lowest terms* when its numerator and denominator are as small as they can be made without changing the value of the fraction. $\frac{1}{2}$ is in its lowest terms; $\frac{2}{4}$ is not.

---

116. What is meant by Reducing a fraction? To what may $\frac{2}{4}$ be reduced? How do you know that $\frac{2}{4}$ and $\frac{1}{2}$ are equal?—117. What operation performed on $\frac{2}{4}$ gives $\frac{1}{2}$? What operation performed on $\frac{1}{2}$ gives $\frac{2}{4}$? By what, then, is the value of a fraction not changed?—118. When is a fraction in its lowest terms?

EXAMPLE.—Reduce $\frac{60}{90}$ to its lowest terms.

Divide both terms by 10, which will make them lower, while it will not change the value of the fraction.

$$10 \,|\, \frac{60}{90} = \frac{6}{9}$$

Looking at the resulting fraction $\frac{6}{9}$, we see that its terms can be divided by 3, which will make them lower without changing its value. Dividing, we get $\frac{2}{3}$.

$$3 \,|\, \frac{6}{9} = \frac{2}{3}$$

As we can find no number greater than 1 that will exactly divide the terms of this fraction, we know that it is in its lowest terms. Answer, $\frac{2}{3}$.

One division may reduce a fraction to its lowest terms, or several may be needed.

119. RULE.—*To reduce a fraction to its lowest terms, divide its numerator and denominator by any number greater than* 1 *that is exactly contained in both. Divide the result in the same way, repeating the process till no number greater than* 1 *is exactly contained in both.*

EXAMPLES FOR THE SLATE.

Reduce the following fractions to their lowest terms :—

1. $\frac{3}{6}$. $\frac{8}{10}$. $\frac{18}{20}$. $\frac{9}{12}$. $\frac{15}{27}$. $\frac{8}{12}$. $\frac{10}{12}$. $\frac{5}{15}$. $\frac{18}{27}$. $\frac{21}{35}$.

2. Reduce $\frac{24}{60}$.	*Ans.* $\frac{2}{5}$.	9. Reduce $\frac{60}{130}$.	*Ans.* $\frac{6}{13}$.
3. Reduce $\frac{11}{55}$.	*Ans.* $\frac{1}{5}$.	10. Reduce $\frac{60}{140}$.	*Ans.* $\frac{3}{7}$.
4. Reduce $\frac{15}{90}$.	*Ans.* $\frac{1}{6}$.	11. Reduce $\frac{225}{275}$.	*Ans.* $\frac{9}{11}$.
5. Reduce $\frac{24}{108}$.	*Ans.* $\frac{2}{9}$.	12. Reduce $\frac{99}{143}$.	*Ans.* $\frac{9}{13}$.
6. Reduce $\frac{80}{112}$.	*Ans.* $\frac{5}{7}$.	13. Reduce $\frac{84}{300}$.	*Ans.* $\frac{7}{25}$.
7. Reduce $\frac{66}{154}$.	*Ans.* $\frac{3}{7}$.	14. Reduce $\frac{32}{80}$.	*Ans.* $\frac{2}{5}$.
8. Reduce $\frac{700}{900}$.	*Ans.* $\frac{7}{9}$.	15. Reduce $\frac{105}{147}$.	*Ans.* $\frac{5}{7}$.

Reduce $\frac{60}{90}$ to its lowest terms. How many divisions are needed, to reduce a fraction to its lowest terms?—**119.** Give the rule for reducing a fraction to its lowest terms.

Reducing Improper Fractions.

120. An Improper Fraction is one whose numerator is equal to or greater than its denominator.

121. An improper fraction may be reduced to a whole or mixed number.

EXAMPLES.—1. Reduce $\frac{18}{9}$ to a whole number.

9 ninths make 1 whole; hence in 18 ninths there are as many wholes as 9 ninths are contained times in 18 ninths, or 2. Answer, 2.

2. Reduce $\frac{20}{9}$ to a mixed number.

9 ninths make 1 whole; hence in 20 ninths there are as many wholes as 9 ninths are contained times in 20 ninths, or $2\frac{2}{9}$. Answer, $2\frac{2}{9}$.

In both these examples, it will be seen, the numerator is divided by the denominator. Hence the rule.

122. RULE.—*To reduce an improper fraction to a whole or mixed number, divide the numerator by the denominator.*

When the numerator is equal to the denominator, the value of the fraction is 1. $\frac{3}{3} = 1$.

When the numerator is greater than the denominator, the value of the fraction is greater than 1. $\frac{4}{3} = 1\frac{1}{3}$.

When, on dividing, a remainder occurs, it must be placed over the denominator in the form of a fraction. The answer is then a mixed number, as in Example 2.

The fraction thus formed, if not already in its lowest terms, must be reduced.

---

120. What is an Improper Fraction?—121. To what may an improper fraction be reduced? Give examples.—122. Give the rule for reducing an improper fraction to a whole or mixed number. When is the value of the fraction 1? When is it greater than 1? When, on dividing, a remainder occurs, what must be done? If this fraction is not already in its lowest terms, what must be done?

EXAMPLE.—Reduce $\frac{40}{26}$ to a whole or mixed number.

Divide the numerator by the denominator.

$$26\ )\ 40\ (\ 1$$
$$\underline{26}$$
$$14$$

Quotient. $1\frac{14}{26}$

Reduce $\frac{14}{26}$ to its lowest terms.

$2)\ \frac{14}{26} = \frac{7}{13}$. *Answer*, $1\frac{7}{13}$.

EXAMPLES FOR THE SLATE.

Reduce these fractions to whole or mixed numbers :—

1. $\frac{6}{5}$. $\frac{11}{9}$. $\frac{15}{11}$. $\frac{12}{7}$. $\frac{7}{4}$. $\frac{16}{15}$. $\frac{11}{6}$. $\frac{8}{5}$.
2. Reduce $\frac{8}{6}$. *Ans.* $1\frac{1}{3}$.
3. Reduce $\frac{30}{8}$. *Ans.* $3\frac{3}{4}$.
4. Reduce $\frac{95}{13}$. *Ans.* $7\frac{4}{13}$.
5. Reduce $\frac{124}{24}$. *Ans.* $5\frac{1}{6}$.
6. Reduce $\frac{180}{19}$. *Ans.* $9\frac{9}{19}$.
7. Reduce $\frac{240}{4}$. *Ans.* 60.
8. Reduce $\frac{354}{32}$. *Ans.* $11\frac{1}{16}$.
9. Reduce $\frac{910}{36}$. *Ans.* $25\frac{5}{18}$.
10. Reduce $\frac{893}{44}$. *Ans.* $20\frac{9}{22}$.
11. Reduce $\frac{966}{26}$. *Ans.* $37\frac{2}{13}$.
12. Reduce $\frac{780}{9}$. *Ans.* $86\frac{2}{3}$.
13. Reduce $\frac{999}{27}$. *Ans.* 37.

Reducing Mixed Numbers to Improper Fractions.

123. A mixed number may be reduced to an improper fraction.

Reduce $12\frac{1}{4}$ to an improper fraction. In 1 there are 4 fourths; and in 12, twelve times 4 fourths, or 48 fourths. 48 fourths and 1 fourth make 49 fourths. Ans. $\frac{49}{4}$.

$$12\frac{1}{4}$$
$$\underline{4}$$
$$48 + 1 = 49$$

Ans. $\frac{49}{4}$.

Hence the following rule :—

124. RULE.—*To reduce a mixed number to an improper fraction, multiply the whole number by the denominator of the fraction, add in the numerator, and set the result over the denominator.*

---

123. To what may a mixed number be reduced? Give an example.—124. What is the rule for reducing a mixed number to an improper fraction?

125. We have seen, in § 115, how to reduce *a whole number* to an improper fraction. The process is the same as that just shown, except that there is no numerator to be added in.

A whole number may also be reduced to a fractional form by giving it 1 for a denominator. $3 = \frac{3}{1}$ $7 = \frac{7}{1}$.

EXAMPLES FOR THE SLATE.

1. Reduce $8\frac{4}{9}$ to an improper fraction. *Ans.* $\frac{76}{9}$.
2. Reduce $4\frac{4}{11}$.
3. Reduce $20\frac{5}{8}$.
4. Reduce $39\frac{1}{7}$.
5. Reduce $25\frac{1}{8}$.
6. Reduce $137\frac{4}{5}$.
7. Reduce $29\frac{7}{16}$.
8. Reduce $47\frac{3}{20}$.
9. Reduce $15\frac{2}{5}$.
10. Reduce 20 to a fractional form. *Ans.* $\frac{20}{1}$.
11. Reduce 15 to ninetieths. *Ans.* $\frac{1350}{90}$.
12. How many quarters in 52 yards?
13. How many twelfths in $43\frac{7}{12}$? In $5\frac{9}{12}$?
14. How many nineteenths in 43? In $8\frac{8}{19}$?
15. Reduce $100\frac{7}{10}$ to an improper fraction.
16. Reduce $999\frac{1}{11}$ to an improper fraction.
17. Reduce $640\frac{3}{27}$ to an improper fraction.

Reducing to a Common Denominator.

126. Fractions are said to have *a common denominator*, when their denominators are the same. $\frac{2}{5}$ and $\frac{4}{5}$ have a common denominator.

127. Fractions whose denominators are different may be reduced to others having a common denominator.

---

125. How may a whole number be reduced to an improper fraction? In what other way may a whole number be reduced to a fractional form? To what whole number is $\frac{9}{1}$ equivalent?—126. When are fractions said to have a common denominator?—127. What may be done with fractions whose denominators are different?

EXAMPLE.—Reduce $\frac{2}{5}$, $\frac{1}{4}$, and $\frac{5}{7}$, to fractions that have a common denominator.

The three denominators are 5, 4, and 7. Now, the product of three factors is the same, in whatever order they are taken. Hence, if we multiply each denominator by the other two, we shall get the same number, and this will be the common denominator.

$$\left.\begin{array}{l} 5 \times 4 \times 7 = 140 \\ 4 \times 5 \times 7 = 140 \\ 7 \times 5 \times 4 = 140 \end{array}\right\} \text{Common denom.}$$

$$\left.\begin{array}{l} \frac{2 \times 4 \times 7}{5 \times 4 \times 7} = \frac{56}{140} \\ \frac{1 \times 5 \times 7}{4 \times 5 \times 7} = \frac{35}{140} \\ \frac{5 \times 5 \times 4}{7 \times 5 \times 4} = \frac{100}{140} \end{array}\right\} \textit{Ans.}$$

But the value of the fractions must not be changed. We must, therefore, multiply each numerator by the same multipliers as its denominator. Hence the following rule:—

128. RULE.—*To reduce fractions to others having a common denominator, multiply both terms of each fraction by all the denominators except its own.*

129. Whole numbers must first be reduced to a fractional form, and mixed numbers to improper fractions.

EXAMPLES FOR THE SLATE.

Reduce the following fractions to equivalent ones having a common denominator:—

1. Reduce $\frac{2}{3}$ and $\frac{9}{14}$. *Ans.* $\frac{28}{42}$, $\frac{27}{42}$.
2. Reduce $\frac{1}{2}$, $\frac{5}{3}$, and $\frac{4}{5}$. *Ans.* $\frac{15}{30}$, $\frac{50}{30}$, $\frac{24}{30}$.
3. Reduce $\frac{3}{10}$, $\frac{5}{7}$, and $\frac{2}{9}$. *Ans.* $\frac{189}{630}$, $\frac{450}{630}$, $\frac{140}{630}$.
4. Reduce $\frac{5}{7}$, $\frac{1}{6}$, and $\frac{10}{11}$. *Ans.* $\frac{330}{462}$, $\frac{77}{462}$, $\frac{420}{462}$.
5. Reduce 3, $\frac{1}{15}$, and $\frac{5}{8}$. *Ans.* $\frac{360}{120}$, $\frac{8}{120}$, $\frac{75}{120}$.
6. Reduce $\frac{7}{10}$, $4\frac{1}{3}$, and 6. *Ans.* $\frac{21}{30}$, $\frac{130}{30}$, $\frac{180}{30}$.

With the given example, show how fractions may be reduced to others having a common denominator.—128. Give the rule.—129. To what must whole numbers first be reduced? To what must mixed numbers first be reduced?

7. Reduce $\frac{3}{4}$, $\frac{1}{5}$, $\frac{7}{3}$, and $\frac{5}{7}$, to fractions having a common denominator. *Ans.* $\frac{315}{420}$, $\frac{84}{420}$, $\frac{980}{420}$, $\frac{300}{420}$.

8. Reduce $\frac{1}{2}$, 1, $\frac{9}{13}$, $\frac{2}{5}$. *Ans.* $\frac{65}{130}$, $\frac{130}{130}$, $\frac{90}{130}$, $\frac{52}{130}$.

9. Reduce $\frac{3}{17}$, $\frac{3}{5}$, $1\frac{1}{6}$. *Ans.* $\frac{90}{510}$, $\frac{306}{510}$, $\frac{595}{510}$.

10. Reduce $\frac{13}{19}$ and $\frac{20}{21}$. *Ans.* $\frac{378}{390}$, $\frac{380}{390}$.

11. Reduce $\frac{1}{3}$, $\frac{5}{7}$, $\frac{3}{2}$, 5. *Ans.* $\frac{14}{42}$, $\frac{30}{42}$, $\frac{63}{42}$, $\frac{210}{42}$.

12. Reduce $\frac{3}{11}$, $2\frac{2}{5}$, $\frac{1}{3}$, $\frac{1}{2}$. *Ans.* $\frac{90}{830}$, $\frac{792}{330}$, $\frac{110}{330}$, $\frac{165}{330}$.

13. Reduce $\frac{1}{12}$, $\frac{5}{7}$, and $\frac{2}{5}$. *Ans.* $\frac{35}{420}$, $\frac{300}{420}$, $\frac{168}{420}$.

14. Reduce $\frac{5}{31}$ and $\frac{7}{20}$. *Ans.* $\frac{100}{620}$, $\frac{217}{620}$.

15. Reduce $3\frac{1}{3}$, $\frac{1}{5}$, and $2\frac{1}{2}$. *Ans.* $\frac{100}{30}$, $\frac{6}{30}$, $\frac{75}{30}$.

Addition of Fractions.

130. Halves and halves, thirds and thirds, &c., can be added, just as we add pears and pears, dollars and dollars.

EXAMPLE.—What is the sum of 5 halves and 4 halves? Answer, 9 halves.

$$\frac{5}{2} + \frac{4}{2} = \frac{9}{2}$$

Here the denominators are the same, and we simply add the numerators, and place the sum over the common denominator.

1. Add $\frac{3}{10}$, $\frac{1}{10}$, and $\frac{5}{10}$.
2. Add $\frac{1}{7}$, $\frac{2}{7}$, and $\frac{6}{7}$.
3. Add $\frac{3}{20}$, $\frac{8}{20}$, and $\frac{6}{20}$.
4. Add $\frac{1}{8}$, $\frac{2}{8}$, and $\frac{10}{8}$.
5. Add $\frac{5}{60}$, $\frac{7}{60}$, and $\frac{9}{60}$.
6. Add $\frac{2}{9}$, $\frac{3}{9}$, $\frac{1}{9}$, and $\frac{4}{9}$.
7. Add $\frac{5}{16}$, $\frac{3}{16}$, and $\frac{7}{16}$.
8. Add $\frac{5}{6}$, $\frac{7}{6}$, $\frac{1}{6}$, and $\frac{8}{6}$.

131. Halves and thirds, halves and fourths, &c., can not be thus *directly* added, any more than we can add pears and dollars. They are things of different kinds.

130. Can halves and halves, thirds and thirds, &c., be added? In what way? Give an example.—131. Can halves and thirds, halves and fourths, &c., be added directly? Why not?

EXAMPLE.—What is the sum of 5 thirds and 3 halves?

The parts being of different value, we can not put them together, and say they make 8 halves or 8 thirds. But, if we reduce them to parts of the same kind or value, we can then add them.

By reducing to a common denominator, we find that 5 thirds are equal to 10 sixths, and 3 halves to 9 sixths. 10 sixths and 9 sixths make 19 sixths. $\frac{19}{6}$ being an improper fraction, we reduce it to $3\frac{1}{6}$. Answer, $3\frac{1}{6}$.

$$\frac{5}{3} = \frac{10}{6}$$
$$\frac{3}{2} = \frac{9}{6}$$
$$\frac{10}{6} + \frac{9}{6} = \frac{19}{6}$$
$$\frac{19}{6} = 3\frac{1}{6} \quad \textit{Ans.}$$

132. RULE.—1. *To add fractions, when they have a common denominator, add their numerators, and place the sum over the common denominator.*

When they have not a common denominator, reduce them to fractions that have, and then proceed as above.

If the resulting fraction is not in its lowest terms, reduce it. If it is an improper fraction, reduce it to a whole or mixed number.

2. *To add mixed numbers, or fractions and whole numbers together, find the sum of the fractions separately, and add it to the sum of the whole numbers.*

EXAMPLE.—Add together $\frac{3}{5}$, $4\frac{1}{2}$, $\frac{2}{3}$, and 5.

Add the fractions: $\frac{3}{5} + \frac{1}{2} + \frac{2}{3} = 1\frac{23}{30}$.

Add the whole numbers: $4 + 5 = 9$.

Add these two sums:

$$\begin{array}{r} 1\frac{23}{30} \\ 9 \\ \hline \textit{Ans. } 10\frac{23}{30} \end{array}$$

Show how we add $\frac{5}{3}$ and $\frac{3}{2}$.—132. Give the rule for adding fractions. Give the rule for adding mixed numbers, or fractions and whole numbers. Apply this latter rule in the given example.

EXAMPLES FOR THE SLATE.

1. Find the sum of $\frac{1}{4}$ and $\frac{5}{6}$. *Ans.* $1\frac{1}{12}$.

2. Find the sum of $\frac{7}{9}$ and $\frac{1}{5}$. *Ans.* $\frac{44}{45}$.

3. Find the sum of $\frac{3}{20}$ and $\frac{5}{3}$. *Ans.* $1\frac{49}{60}$.

4. Add together $\frac{1}{2}$, $1\frac{2}{5}$, and $\frac{7}{8}$. *Ans.* $2\frac{31}{40}$.

5. Add together $\frac{3}{20}$ and $2\frac{7}{27}$. *Ans.* $2\frac{221}{540}$.

6. Add together $\frac{2}{9}$, $\frac{1}{10}$, and $\frac{3}{11}$. *Ans.* $\frac{589}{990}$.

7. Add together $\frac{4}{25}$, 10, and $8\frac{7}{9}$. *Ans.* $18\frac{211}{225}$.

8. Add together $7\frac{1}{2}$, $\frac{5}{9}$, 4, and $\frac{1}{5}$. *Ans.* $12\frac{23}{90}$.

9. Add together 3, $4\frac{1}{3}$, $\frac{2}{7}$, and $1\frac{1}{2}$. *Ans.* $9\frac{5}{42}$.

10. Add together $\frac{5}{6}$, $\frac{2}{7}$, $\frac{3}{4}$, and $2\frac{2}{3}$. *Ans.* $4\frac{45}{84}$.

11. A hackman earns \$$2\frac{1}{4}$ one day, \$$3\frac{1}{2}$ the next, \$4 the next, and \$$5\frac{1}{4}$ the next. How much does he earn in all four days? *Ans.* \$15.

12. How much land is there in 3 fields, containing $14\frac{1}{2}$, $7\frac{1}{10}$, and $23\frac{2}{5}$ acres? *Ans.* 45 acres.

13. If I buy \$$2\frac{3}{4}$ worth of paper, and \$$6\frac{1}{4}$ worth of books, and give the storekeeper a ten-dollar bill, how much change will I receive? *Ans.* \$1.

14. Three men, buying a meadow, put in respectively \$$30\frac{1}{10}$, \$$25\frac{2}{10}$, and \$$19\frac{7}{10}$. What does the meadow cost?

15. A five-pound jar contains $3\frac{2}{3}$ pounds of bread and $2\frac{4}{9}$ pounds of cake; what does the whole weigh?

16. A peddler walks $8\frac{1}{8}$ miles one day, $5\frac{1}{4}$ the next, $10\frac{1}{2}$ the next, and 12 the next; how far does he walk in all?

17. How many pecks of peaches in four baskets, containing respectively $2\frac{1}{2}$, $3\frac{1}{3}$, $2\frac{3}{4}$, and $3\frac{1}{6}$ pecks?

18. If $5\frac{6}{7}$ gallons of brandy are mixed with $1\frac{9}{10}$ gallons of water, how many gallons are there of the mixture?

19. A lady hires a gardener for 15 cents an hour. How much must she pay him, if he works $6\frac{5}{12}$ hours the first day, $7\frac{5}{6}$ the second, and $5\frac{3}{4}$ the third? *Ans.* 300 cents.

Subtraction of Fractions.

133. The same principle applies in subtracting, as in adding, fractions. Before subtracting, the parts must be made of the same kind or value, if they are not already so; that is, the denominators must be made the same.

EXAMPLES.—1. From 5 halves take 4 halves. Answer, 1 half. $\frac{5}{2} - \frac{4}{2} = \frac{1}{2}$

2. From 5 thirds take 3 halves.

Thirds and halves being parts of different value, we must reduce them to parts of the same value. Reducing to a common denominator, we find that 5 thirds are equal to 10 sixths, and 3 halves to 9 sixths. 9 sixths from 10 sixths leave 1 sixth. Answer, $\frac{1}{6}$.

$$\frac{5}{3} = \frac{10}{6}$$
$$\frac{3}{2} = \frac{9}{6}$$
$$\frac{10}{6} - \frac{9}{6} = \frac{1}{6}$$
Ans. $\frac{1}{6}$

134. RULE.—1. *To subtract one fraction from another, when they have a common denominator, take the numerator of the subtrahend from that of the minuend, and place the remainder over the common denominator.*

When they have not a common denominator, reduce them to fractions that have, and then proceed as above.

Reduce the resulting fraction to its lowest terms, or to a whole or mixed number, as may be necessary.

2. *Whole and mixed numbers may be reduced to improper fractions, before subtracting.*

---

133. What principle applies in subtracting fractions? Illustrate this, with the examples given.—134. Recite the rule. What may be done with whole and mixed numbers, before subtracting?

EXAMPLES FOR THE SLATE.

1. From $\frac{6}{7}$ take $\frac{4}{7}$.
2. From $\frac{14}{19}$ take $\frac{3}{19}$.
3. Take $\frac{7}{20}$ from $\frac{11}{20}$.
4. From $\frac{19}{42}$ take $\frac{5}{42}$.
5. From $\frac{21}{100}$ take $\frac{1}{100}$.
6. Take $\frac{29}{50}$ from $\frac{49}{50}$.
7. Subtract $\frac{5}{6}$ from $\frac{19}{20}$. *Ans.* $\frac{7}{60}$.
8. From $\frac{3}{11}$ subtract $\frac{1}{4}$. *Ans.* $\frac{1}{44}$.
9. Subtract $\frac{5}{18}$ from $\frac{4}{3}$. *Ans.* $1\frac{1}{18}$.
10. From $\frac{38}{6}$ subtract $\frac{12}{9}$. *Ans.* 5.
11. Take $\frac{1}{10}$ from $\frac{2}{3}$.
12. Take $\frac{3}{100}$ from $\frac{5}{7}$.
13. Take $\frac{2}{19}$ from $\frac{1}{9}$.
14. From $\frac{1}{6}$ take $\frac{1}{7}$.
15. From $\frac{21}{5}$ take $\frac{5}{4}$.
16. From $\frac{19}{8}$ take $\frac{6}{16}$.
17. From 1 subtract $\frac{2}{3}$. *Ans.* $\frac{1}{3}$.
    [Reduce 1 to thirds; then proceed as before.]
18. From 1 take $\frac{17}{20}$. From 1 take $\frac{3}{40}$.
19. From 4 subtract $\frac{2}{3}$. *Ans.* $3\frac{1}{3}$.
    [Take 1 of the 4 units, and reduce it to thirds. Then subtract the fraction. $\frac{2}{3}$ from $\frac{3}{3}$ leaves $\frac{1}{3}$. Ans. $3\frac{1}{3}$.
    $4 = 3\frac{3}{3}$ Min.
    $\frac{2}{3}$ Sub.
    $3\frac{1}{3}$ Rem.]
20. From 5 subtract $\frac{3}{7}$. [$5 = 4\frac{7}{7}$] *Ans.* $4\frac{4}{7}$.
21. From 17 subtract $\frac{9}{14}$. *Ans.* $16\frac{5}{14}$.
22. Take $\frac{1}{10}$ from 3. Take $\frac{5}{12}$ from 3. Take $\frac{2}{7}$ from 2.
23. Subtract $\frac{1}{3}$ from $5\frac{3}{4}$. *Ans.* $5\frac{5}{12}$.
    [Take $\frac{1}{3}$ from $\frac{3}{4}$, and annex the result to 5.]
24. Subtract $\frac{1}{2}$ from $2\frac{1}{2}$. From $2\frac{3}{4}$. Take $\frac{2}{3}$ from $6\frac{4}{5}$.
25. Subtract $\frac{5}{8}$ from $6\frac{7}{9}$. *Ans.* $6\frac{11}{72}$.
26. Subtract $2\frac{1}{5}$ from $4\frac{1}{6}$. *Ans.* $1\frac{29}{30}$.
    [$\frac{1}{5}$ is greater than $\frac{1}{6}$. Hence reduce both mixed numbers to improper fractions. The sum then becomes, Subtract $\frac{11}{5}$ from $\frac{25}{6}$. Now proceed as before.]
27. Subtract $1\frac{3}{10}$ from $3\frac{2}{15}$. *Ans.* $1\frac{5}{6}$.
28. Subtract $20\frac{1}{2}$ from $24\frac{1}{3}$. *Ans.* $3\frac{5}{6}$.
29. Subtract $\frac{97}{100}$ from 12. *Ans.* $11\frac{3}{100}$.
30. Subtract $\frac{8}{9}$ from $1\frac{2}{9}$. *Ans.* $\frac{1}{3}$.

31. Subtract $1\frac{4}{5}$ from $9\frac{1}{2}$. *Ans.* $7\frac{7}{10}$.

32. A grocer, having mixed $14\frac{1}{3}$ pounds of tea with $26\frac{4}{6}$ pounds of a different kind, sold all the mixture but 11 pounds. How much did he sell?

33. If a person owning two farms, one of $70\frac{3}{5}$ acres, and the other of $120\frac{3}{10}$ acres, sells $90\frac{7}{10}$ acres, how much land has he remaining? *Ans.* $100\frac{1}{5}$ acres.

34. How much paper has a printer left, if he had on hand $27\frac{1}{2}$ reams, and has used $7\frac{1}{4}$ reams for one job, and $6\frac{4}{5}$ reams for another? *Ans.* $13\frac{9}{20}$ reams.

Multiplication of Fractions.

Fraction × Whole Number.

135. To multiply any number of equal parts by 2, we may either take twice as many such parts, or make each part twice as great. That is, we may double the *number* of parts, or double their *size*.

EXAMPLE.—Multiply $\frac{3}{4}$ by 2.

Double the *number* of parts. Twice $\frac{3}{4}$ is $\frac{6}{4}$.

Or, double the *size* of the parts. A half is twice as great as a fourth. Hence, twice $\frac{3}{4}$ is $\frac{3}{2}$.

$\frac{6}{4}$ may be reduced to $\frac{3}{2}$. The two answers agree.

Now, in the first case, we multiplied the numerator by 2. In the second case, we divided the denominator by 2. The latter mode is better, because it brings the fraction at once in its lowest terms. Hence the rule.

136. RULE.—*To multiply a fraction by a whole number, divide its denominator by the whole number, if this can be done without a remainder; if not, multiply its numerator.*

135. To multiply any number of equal parts by 2, what may we do? Illustrate these two methods, in multiplying $\frac{3}{4}$ by 2. Which method is better?—136. Give the rule for multiplying a fraction by a whole number.

EXAMPLES.—1. Multiply $\frac{3}{16}$ by 4.
16 can be divided by 4. Divide it. Answer, $\frac{3}{4}$.

2. Multiply $\frac{3}{17}$ by 4.
17 can not be divided by 4 without a remainder. Multiply the numerator. Answer, $\frac{12}{17}$.

137. Find the value of the following:—

1. $\frac{2}{3} \times 7$.	*Ans.* $4\frac{2}{3}$.	7. $\frac{14}{121} \times 11$.	*Ans.* $1\frac{3}{11}$.
2. $\frac{5}{6} \times 2$.	*Ans.* $1\frac{2}{3}$.	8. $\frac{3}{21} \times 4$.	*Ans.* $\frac{12}{21}$.
3. $\frac{7}{12} \times 4$.	*Ans.* $2\frac{1}{3}$.	9. $\frac{9}{108} \times 12$.	*Ans.* 1.
4. $\frac{3}{25} \times 5$.	*Ans.* $\frac{3}{5}$.	10. $\frac{17}{156} \times 12$.	*Ans.* $1\frac{4}{13}$.
5. $\frac{2}{13} \times 6$.		11. $\frac{45}{64} \times 8$.	
6. $\frac{1}{36} \times 6$.		12. $\frac{11}{100} \times 10$.	

MIXED NUMBER × WHOLE NUMBER.

138. RULE.—*To multiply a mixed number by a whole number, multiply the fractional part and the whole part of the mixed number separately, and add the products.*

EXAMPLE.—Multiply $20\frac{3}{10}$ by 5.

Multiply the fraction: $\frac{3}{10} \times 5 = \frac{3}{2} = 1\frac{1}{2}$
Multiply the whole part: $20 \times 5 = 100$
Add the products: $101\frac{1}{2}$ *Ans.*

1. What will 10 clocks cost, at $\$5\frac{1}{5}$ each? *Ans.* $52.

2. There are $5\frac{1}{2}$ yards in a rod; how many yards are there in 40 rods? *Ans.* 220 yards.

3. If 12 acres of land produce, on an average, $35\frac{1}{3}$ bushels of wheat each, what is the yield of the whole?

4. A locomotive is moving at the rate of $25\frac{7}{8}$ miles an hour; how far will it go in 3 hours? *Ans.* $77\frac{5}{8}$ miles.

5. What will 16 chairs cost, at $\$2\frac{3}{4}$ apiece?

---

138. Give the rule for multiplying a mixed number by a whole number.

WHOLE NUMBER × FRACTION.

139. Multiplying a number by a fraction is simply taking such a part of it as is denoted by the fraction.

Multiplying a number by $\frac{1}{3}$ is taking $\frac{1}{3}$ of it, or dividing it by 3, as shown in § 112. Multiplying by $\frac{2}{3}$ is taking $\frac{1}{3}$ twice, or dividing by the denominator 3, and multiplying by the numerator 2, as shown in § 113.

In such cases, it is best to multiply first, as there may be a remainder on dividing. Hence the rule.

140. RULE.—*To multiply a whole number by a fraction, multiply it by the numerator of the fraction, and divide by the denominator.*

EXAMPLE.—Multiply 140 by $\frac{2}{3}$.

Multiply 140 by the numerator 2, and divide the product by the denominator 3. Answer, $93\frac{1}{3}$.

```
   140
     2
3)280
  93⅓ Ans.
```

1. Multiply 160 by $\frac{2}{3}$.
2. Multiply 199 by $\frac{4}{5}$.
3. Multiply 200 by $\frac{1}{6}$.
4. $5713 \times \frac{5}{7}$. *Ans.* $4080\frac{5}{7}$.
5. $8296 \times \frac{7}{9}$. *Ans.* $6452\frac{4}{9}$.
6. $2644 \times \frac{10}{11}$. *Ans.* $2403\frac{7}{11}$.
7. A farmer owning 467 acres of land, sells $\frac{3}{4}$ of it; how many acres does he sell? *Ans.* $350\frac{1}{4}$ acres.
8. How many pounds in $\frac{11}{15}$ of a ton of coal, a ton being 2000 pounds? *Ans.* $1466\frac{2}{3}$ pounds.
9. How many pounds in $\frac{5}{8}$ of a barrel of flour, there being 196 pounds in a barrel?
10. If a hogshead holds 63 gallons, and $\frac{2}{21}$ of its contents has leaked out, how much remains? *Ans.* 57 gallons.

---

139. To what is multiplying a number by a fraction equivalent? To what, for instance, is multiplying by $\frac{1}{3}$ equivalent? By $\frac{2}{3}$?—140. Give the rule for multiplying a whole number by a fraction.

8

Whole Number × Mixed Number.

141. Rule.—*To multiply a whole number by a mixed number, multiply it first by the fractional part of the mixed number, then by the whole part, and add the products.*

	317
Example.—Multiply 317 by $6\frac{2}{3}$.	$6\frac{2}{3}$
Multiply 317 by $\frac{2}{3}$:	$211\frac{1}{3}$
Multiply 317 by 6:	1902
Add the products:	$2113\frac{1}{3}$ *Ans.*

1. Multiply 2463 by $7\frac{1}{2}$. *Ans.* $18472\frac{1}{2}$.
2. Multiply 5698 by $6\frac{3}{5}$. *Ans.* $37606\frac{4}{5}$.
3. Multiply 1275 by $39\frac{4}{15}$. *Ans.* 50065.
4. Multiply 5416 by $85\frac{3}{20}$. *Ans.* $461172\frac{2}{5}$.
5. Multiply 8570 by $73\frac{1}{4}$. *Ans.* $627752\frac{1}{2}$.
6. Multiply 9087 by $20\frac{10}{11}$. *Ans.* $190000\frac{10}{11}$.
7. Multiply 2639 by $62\frac{5}{13}$. *Ans.* 164633.
8. Multiply 4471 by $49\frac{7}{17}$. *Ans.* 220920.
9. Multiply 6381 by $56\frac{4}{9}$. *Ans.* 360172.

Fraction × Fraction.

142. Multiplying by a fraction, we have seen, is equivalent to taking such a part as is denoted by the fraction. Multiplying $\frac{1}{2}$ by $\frac{1}{4}$ is equivalent to taking $\frac{1}{4}$ of $\frac{1}{2}$.

143. A fraction of a fraction, such as $\frac{1}{4}$ of $\frac{1}{2}$, is called a **Compound Fraction.**

The same process is used in multiplying fractions and in reducing compound fractions to simple ones.

---

141. Give the rule for multiplying a whole number by a mixed number.
142. To what is multiplying by a fraction equivalent?—143. What is a fraction of a fraction called? In what two operations is the same process used?

144. Multiply $\frac{1}{2}$ by $\frac{3}{4}$.

These fractions indicate division. The numerators are the dividends. The denominators are the divisors. Multiply the numerators together to find the total dividend, and the denominators to find the total divisor.
Then set the former product over the latter, in the form of a fraction.

$\frac{1}{2} \times \frac{3}{4} = \frac{3}{8}$ *Ans.*

1. Multiply $\frac{5}{6}$ by $\frac{7}{8}$. *Ans.* $\frac{35}{48}$.
2. Multiply $\frac{1}{3}$, $\frac{1}{4}$, and $\frac{7}{9}$ together. *Ans.* $\frac{7}{108}$.
3. Multiply $\frac{2}{5}$, $\frac{4}{7}$, and $\frac{6}{11}$ together. *Ans.* $\frac{48}{385}$.
4. Multiply $\frac{2}{3}$, $\frac{4}{5}$, $\frac{1}{7}$, and $\frac{1}{3}$ together. *Ans.* $\frac{8}{315}$.
5. Reduce $\frac{2}{5}$ of $\frac{1}{6}$ of $\frac{7}{9}$ to a simple fraction. *Ans.* $\frac{14}{270}$.

145. Cancelling.—When the same number appears as a numerator and a denominator, draw a line through it, and omit it in multiplying. This is called **Cancelling.**

Ex.—Reduce $\frac{2}{3}$ of $\frac{5}{9}$ of $\frac{4}{5}$ of $\frac{3}{9}$ to a simple fraction.

$$\frac{2}{\cancel{3}} \text{ of } \frac{\cancel{5}}{9} \text{ of } \frac{4}{\cancel{5}} \text{ of } \frac{\cancel{3}}{9} = \frac{8}{81} \ \textit{Ans.}$$

Cancel 3 and 3, 5 and 5. Then multiply the remaining numerators and denominators. Observe that 9 and 9 are not cancelled, because they are both denominators.

1. Multiply $\frac{5}{6}$ by $\frac{6}{7}$. *Ans.* $\frac{5}{7}$.
2. Multiply $\frac{3}{4}$, $\frac{1}{3}$, and $\frac{4}{5}$ together. *Ans.* $\frac{1}{5}$.
3. Multiply $\frac{7}{8}$, $\frac{3}{11}$, $\frac{11}{7}$, and $\frac{3}{5}$ together. *Ans.* $\frac{3}{5}$.
4. Reduce $\frac{5}{9}$ of $\frac{2}{3}$ of $\frac{9}{13}$ to a simple fraction. *Ans.* $\frac{10}{39}$.
5. Reduce $\frac{7}{8}$ of $\frac{5}{11}$ of $\frac{3}{7}$ to a simple fraction. *Ans.* $\frac{15}{88}$.
6. Reduce $\frac{5}{21}$ of $\frac{2}{5}$ of $\frac{21}{3}$ to a simple fraction.
7. Reduce $\frac{6}{7}$ of $\frac{2}{3}$ of $\frac{7}{2}$ of $\frac{1}{6}$ to a simple fraction.

---

144. In multiplying $\frac{1}{2}$ by $\frac{3}{4}$, show how we proceed, and why.—145. When the same number appears as a numerator and a denominator, what do we do? What is this called? Give an example.

146. Equal factors in a numerator and a denominator may also be cancelled.

Ex.—Reduce $\frac{2}{9}$ of $\frac{6}{7}$ of 35 to a simple fraction.

Throw the whole number 35 into a fractional form.

$$\frac{2}{\overset{}{\underset{3}{\not 9}}} \text{ of } \frac{\overset{2}{\not 6}}{\not 7} \text{ of } \frac{\overset{5}{\not 3\not 5}}{1} = \frac{20}{3} = 6\tfrac{2}{3} \textit{ Ans.}$$

Cancel 3 (that is, divide by 3) in the first denominator and second numerator. Cancel 7 (that is, divide by 7) in the second denominator and third numerator. Then multiply the remaining factors.

1. Multiply $\frac{3}{8}$, $\frac{4}{7}$, and $\frac{9}{5}$ together. *Ans.* $\frac{27}{70}$.
2. Multiply $\frac{1}{3}$, $\frac{6}{7}$, and 14 together. *Ans.* 4.
3. Multiply $\frac{5}{6}$, $\frac{9}{4}$, $\frac{3}{5}$, and $\frac{3}{7}$ together. *Ans.* $\frac{27}{56}$.
4. Reduce $\frac{6}{11}$ of $\frac{7}{3}$ of $\frac{22}{9}$ of $\frac{1}{3}$. *Ans.* $\frac{28}{45}$.

147. Cancelling is dividing. When, therefore, a numerator or denominator disappears by cancelling, 1 (*not* 0) is left in its place.

$$\frac{\not 3}{\not 4} \times \frac{\not 4}{\not 3} = \frac{1}{1} = 1$$

148. Cancelling is dividing. Hence, by cancelling before we multiply, we save the trouble of dividing after we multiply, to reduce the result to its lowest terms.

149. General Rule.—1. *To multiply one fraction by another, or to reduce a compound fraction, first cancel factors common to any numerator and denominator; then multiply the numerators together for a new numerator, and the denominators for a new denominator.*

---

146. In what other way may we cancel? Give an example.—147. When we cancel, what operation do we perform? When a numerator or denominator disappears by cancelling, what is left?—148. What trouble do we avoid, by cancelling before we multiply?—149. Give the general rule for the multiplication of fractions.

2. *Whole numbers occurring in a compound fraction must first be reduced to a fractional form, and mixed numbers to improper fractions.*

EXAMPLES FOR THE SLATE.

1. Multiply together $\frac{3}{10}$, $\frac{6}{7}$, 14, and $\frac{10}{11}$. *Ans.* $3\frac{3}{11}$.
2. Multiply together $\frac{2}{3}$, $\frac{2}{5}$, $\frac{1}{4}$, and 15. *Ans.* 1.
3. Multiply together $\frac{3}{4}$, $\frac{5}{20}$, $\frac{4}{5}$, and $\frac{1}{7}$. *Ans.* $\frac{3}{140}$.
4. Find the product of $\frac{1}{9}$, $\frac{8}{15}$, and $3\frac{15}{16}$. *Ans.* $\frac{7}{30}$.
5. Find the product of $\frac{5}{24}$, $\frac{9}{14}$, $\frac{36}{7}$, and $\frac{1}{10}$. *Ans.* $\frac{27}{392}$.
6. Reduce $\frac{9}{7}$ of $\frac{1}{2}$ of $\frac{5}{3}$ of $1\frac{3}{4}$. *Ans.* $1\frac{7}{8}$.
7. Reduce $\frac{1}{2}$ of $\frac{1}{3}$ of $\frac{1}{5}$ of $\frac{1}{6}$. *Ans.* $\frac{1}{180}$.
8. Reduce $\frac{2}{9}$ of $\frac{5}{3}$ of $\frac{3}{8}$ of 9. *Ans.* $1\frac{1}{4}$.
9. Reduce $\frac{5}{16}$ of $\frac{8}{9}$ of $\frac{1}{5}$ of $4\frac{1}{4}$. *Ans.* $\frac{17}{72}$.
10. Reduce $\frac{5}{6}$ of $\frac{7}{9}$ of $\frac{10}{11}$ of 12. *Ans.* $7\frac{7}{99}$.

MIXED NUMBER × MIXED NUMBER.

150. RULE.—*To multiply two or more mixed numbers together, reduce them to improper fractions, and proceed as in multiplication of fractions.*

EXAMPLE.—Multiply $4\frac{2}{3}$, $5\frac{1}{14}$, and $1\frac{1}{2}$ together.

Reduce the mixed numbers to improper fractions. Then, cancelling, we get $\frac{71}{2}$, or $35\frac{1}{2}$.

$$\frac{\cancel{14}}{\cancel{3}} \times \frac{71}{\cancel{14}} \times \frac{\cancel{3}}{2} = \frac{71}{2} = 35\tfrac{1}{2}\ \textit{Ans.}$$

Find the value of the following :—

1. $26\frac{2}{3} \times 3\frac{1}{2}$. *Ans.* $93\frac{1}{3}$.
2. $7\frac{1}{9} \times 1\frac{7}{8}$. *Ans.* $13\frac{1}{3}$.
3. $1\frac{3}{7} \times 9\frac{3}{5}$. *Ans.* $13\frac{5}{7}$.
4. $3\frac{3}{7} \times 4\frac{2}{3} \times 1\frac{1}{9}$. *Ans.* $17\frac{7}{9}$.
5. $11\frac{1}{5} \times 1\frac{7}{8} \times 1\frac{2}{11}$. *Ans.* $24\frac{9}{11}$.
6. $2\frac{1}{4} \times 3\frac{2}{3} \times 5\frac{2}{5}$. *Ans.* $44\frac{11}{20}$.

---

150. Recite the rule for multiplying two or more mixed numbers together.

EXAMPLES FOR THE SLATE.

1. A merchant owning $\frac{3}{4}$ of a ship, sold $\frac{7}{10}$ of his share. What part of the ship did he sell? *Ans.* $\frac{21}{40}$.

2. What will $15\frac{7}{8}$ yards of velvet cost, at $4\frac{1}{4}$ dollars a yard? *Ans.* \$$67\frac{15}{32}$.

3. A person having $423\frac{1}{3}$ acres of land, left $\frac{3}{5}$ of it to his son. What was the son's share? *Ans.* 254 acres.

4. A farmer has three wheat fields, of $4\frac{1}{5}$ acres each. Their average yield is $33\frac{3}{4}$ bushels to the acre. What is the yield of the whole? *Ans.* $425\frac{1}{4}$ bu.

5. General Putnam lived to be 72 years old. Patrick Henry attained $\frac{7}{8}$ of that age; how old was he at the time of his death?

6. How much flour must be laid in for a garrison of 355 men, to allow each man $56\frac{3}{4}$ pounds?

7. The British House of Commons contains 654 members. $\frac{248}{327}$ of this number are from England and Wales; how many does that make? *Ans.* 496 members.

8. How many yards are there in a bale of linen, containing 56 pieces, if there are $25\frac{2}{3}$ yards in each piece?

9. If a clock ticks sixty times in a minute, how many times will it tick in $15\frac{3}{5}$ hours, there being sixty minutes in an hour? *Ans.* 56160 times.

10. If 680 persons subscribe for a work in three volumes, costing half a guinea a volume, what is the whole amount of the subscription?

11. A owns $\frac{2}{3}$ of a factory. He sells half his share to B, who in turn sells $\frac{5}{6}$ of his share to C. What part of the factory belongs to C? *Ans.* $\frac{5}{18}$.

12. What will be the cost of three boxes of oranges, allowing 96 oranges to the box, at $1\frac{5}{8}$ cents apiece?

13. Multiply $5\frac{3}{7}$, $5\frac{1}{4}$, and $\frac{1}{19}$ together. *Ans.* $1\frac{1}{2}$.

Division of Fractions.

FRACTION ÷ WHOLE NUMBER.

151. To divide any number of equal parts by 2, we may either take half as many such parts, or make each part half as great.

EXAMPLE.—Divide $\frac{4}{5}$ by 2.

Take half the number of parts. Half of $\frac{4}{5}$ is $\frac{2}{5}$.

Or, make each part half as great. A tenth is half as great as a fifth. Hence half of $\frac{4}{5}$ is $\frac{4}{10}$.

$\frac{4}{10}$ can be reduced to $\frac{2}{5}$. The two answers agree.

Now, in the first case, we divided the numerator by 2. In the second case, we multiplied the denominator by 2. The former mode is better, because it brings the fraction at once in its lowest terms. Hence the following rule.

152. RULE.—1. *To divide a fraction by a whole number, divide its numerator by the whole number, if this can be done without a remainder; if not, multiply its denominator.*

2. *To divide a mixed number by a whole number, reduce the mixed number to an improper fraction; then proceed as above.*

EXAMPLES.—1. Divide $\frac{5}{16}$ by 6.

If the numerator 5 contained 6 exactly, we should divide it by 6. As it does not, we multiply the denominator.

$\frac{5}{16 \times 6} = \frac{5}{96}$ *Ans.*

2. Divide $2\frac{4}{7}$ by 6.

Reduce $2\frac{4}{7}$ to an improper fraction, $\frac{18}{7}$. As the numerator 18 contains 6 exactly, divide it by 6.

$\frac{18 \div 6}{7} = \frac{3}{7}$ *Ans.*

---

151. To divide any number of equal parts by 2, what may we do? Illustrate these two methods, in dividing $\frac{4}{5}$ by 2. Which method is better?—152. Give the rule for dividing a fraction by a whole number.

153. Find the value of the following:—

1. $\frac{28}{31} \div 7$. *Ans.* $\frac{4}{31}$.
2. $\frac{63}{100} \div 9$.
3. $\frac{121}{144} \div 11$.
4. $7\frac{1}{7} \div 10$. *Ans.* $\frac{5}{7}$.
5. $2\frac{2}{9} \div 4$.
6. $3\frac{1}{4} \div 6$.
7. $5\frac{1}{5} \div 17$.
8. $\frac{5}{6} \div 9$. *Ans.* $\frac{5}{54}$.
9. $\frac{5}{12} \div 11$.
10. $\frac{4}{15} \div 5$.
11. $1\frac{5}{6} \div 13$. *Ans.* $\frac{11}{78}$.
12. $2\frac{2}{3} \div 3$.
13. $5\frac{1}{9} \div 23$.
14. $5\frac{5}{7} \div 40$.

Fraction ÷ Fraction.

154. Divide $\frac{3}{5}$ by $\frac{2}{7}$.

That is, find how many times $\frac{2}{7}$ is contained in $\frac{3}{5}$. $\frac{1}{7}$ is contained in 1, 7 times. Hence, in $\frac{3}{5}$ it is contained $\frac{3}{5}$ of 7 times, or $\frac{21}{5}$ times.

But $\frac{2}{7}$ is twice as great as $\frac{1}{7}$, and hence is contained only half as many times. $\frac{1}{2}$ of $\frac{21}{5}$ is $\frac{21}{10}$. Answer, $\frac{21}{10}$, or $2\frac{1}{10}$.

Now, what have we done to the dividend $\frac{3}{5}$, to produce $\frac{21}{10}$? We have multiplied its numerator by the denominator of the divisor $\frac{2}{7}$, and multiplied its denominator by the numerator of the divisor. Or, in other words, we have inverted the divisor, and then multiplied the fractions. Hence the rule.

$$\frac{3}{5} \times \frac{7}{2} = \frac{21}{10}$$

155. Rule.—1. *To divide one fraction by another, multiply the dividend by the divisor inverted.*

2. *Whole and mixed numbers must first be reduced to improper fractions.*

Example.—Divide $4\frac{1}{5}$ by $2\frac{1}{3}$.

Reduce the mixed numbers: $\frac{21}{5} \div \frac{7}{3}$

Invert the divisor; cancel equal factors; multiply. $\frac{\overset{3}{\cancel{21}}}{5} \times \frac{3}{\cancel{7}} = \frac{9}{5} = 1\frac{4}{5}$ *Ans.*

---

154. Divide $\frac{3}{5}$ by $\frac{2}{7}$. What have we done to the dividend $\frac{3}{5}$, to produce this result?—155. Give the rule for dividing one fraction by another.

EXAMPLES FOR THE SLATE.

156. Find the value of the following:—

1. $\frac{2}{3} \div \frac{5}{6}$.	*Ans.* $\frac{4}{5}$.	7. $4\frac{1}{2} \div 3\frac{3}{5}$.	*Ans.* $1\frac{1}{4}$.
2. $\frac{3}{10} \div \frac{7}{20}$.	*Ans.* $\frac{6}{7}$.	8. $7\frac{1}{5} \div \frac{12}{25}$.	*Ans.* 15.
3. $\frac{4}{5} \div \frac{2}{15}$.	*Ans.* 6.	9. $9\frac{1}{11} \div 1\frac{3}{7}$.	*Ans.* $6\frac{4}{11}$.
4. $6 \div \frac{3}{4}$.	*Ans.* 8.	10. $\frac{14}{15} \div 2\frac{5}{6}$.	*Ans.* $\frac{28}{85}$.
5. $9 \div \frac{1}{3}$.		11. $5\frac{1}{9} \div \frac{8}{9}$.	
6. $\frac{1}{7} \div \frac{1}{8}$.		12. $\frac{11}{12} \div 3\frac{1}{4}$.	

13. How many clocks, at \$$5\frac{3}{8}$ apiece, can be bought for \$$21\frac{1}{2}$? *Ans.* 4 clocks.

14. If it takes $6\frac{2}{3}$ yards of lace to trim one dress, how many dresses will 60 yards trim? *Ans.* 9 dresses.

15. A flock of sheep yield $104\frac{1}{7}$ pounds of wool. How many sheep are there, if they average $3\frac{6}{7}$ pounds of wool each? *Ans.* 27 sheep.

16. If a locomotive goes $156\frac{1}{2}$ miles in $5\frac{2}{3}$ hours, what is its rate per hour? *Ans.* $27\frac{21}{34}$ miles.

17. If a cow is allowed $\frac{1}{8}$ of a bushel of turnips a day, how long will half a bushel last her?

18. Divide $8\frac{3}{14}$ by $2\frac{1}{7}$. *Ans.* $3\frac{5}{6}$.

MISCELLANEOUS QUESTIONS ON FRACTIONS.—What operation does a fraction indicate? What part of the fraction corresponds with the divisor? What corresponds with the dividend? What corresponds with the quotient? *Ans.* The value of the fraction. Which is greater, $\frac{1}{7}$ or $\frac{1}{8}$? When we increase the denominator of a fraction, do we increase or diminish its value? Which is greater, $\frac{1}{7}$ or $\frac{2}{7}$? When we increase the numerator of a fraction, do we increase or diminish its value? What kind of a fraction is $\frac{4}{4}$? What is its value? Is the value of a proper fraction greater or less than 1? Does multiplying a number by $\frac{1}{2}$ increase or diminish it? Does dividing a number by $\frac{1}{2}$ increase or diminish it? When we take $\frac{1}{2}$ of a number, do we multiply or divide by $\frac{1}{2}$? To what is cancelling equivalent? When can we cancel?

MISCELLANEOUS EXAMPLES.

1. Find the sum, then the difference, then the product, of $\frac{1}{2}$ and $\frac{1}{10}$. Divide $\frac{1}{2}$ by $\frac{1}{10}$.

2. If a boy's wages are $4\frac{1}{4}$ dollars a week, what will they amount to in 52 weeks? *Ans.* \$221.

3. Allowing 240 pins to a paper, how many pins are there in $\frac{5}{12}$ of a paper?

4. Three boys agreed to share their earnings for one week equally. The first earned \$$5\frac{2}{5}$; the second, \$$4\frac{3}{8}$; the third, \$3. What was the share of each? *Ans.* \$$4\frac{31}{120}$.

5. A planter who has 56 hogsheads of sugar, sells $\frac{1}{4}$ of them to one merchant, and $\frac{3}{8}$ to another. How many hogsheads has he left? *Ans.* 21 hogsheads.

6. If a man has to make a journey of $175\frac{3}{10}$ miles, how far will he have to go when he has travelled $48\frac{4}{5}$ miles? *Ans.* $126\frac{1}{2}$ miles.

7. How many quarter-dollars are there in \$1250?

8. A person owning 200 acres of land, leaves $\frac{1}{2}$ of it to his wife, and she divides her portion equally among her 5 sons. What fraction of the whole does each son get, and how many acres? *Ans.* $\frac{1}{10}$, 20 acres.

9. If a merchant sells three dresses, of $10\frac{3}{4}$ yards each, from a piece of calico containing 40 yards, how many yards will he have left? *Ans.* $7\frac{3}{4}$ yards.

10. If a farm of $143\frac{1}{4}$ acres is sold for \$$2575\frac{1}{2}$, what is the price per acre? *Ans.* \$$17\frac{187}{191}$.

11. Three partners buy some silks for \$$1250\frac{1}{2}$, and sell them for \$$1325\frac{7}{8}$. What profit has each? *Ans.* \$$25\frac{1}{8}$.

12. A tailor sold a coat for \$$20\frac{1}{2}$. If the materials cost him \$$9\frac{2}{3}$, and he paid \$$8\frac{1}{8}$ for making it, what was his profit? *Ans.* \$$2\frac{17}{24}$.

FEDERAL MONEY.

157. Different countries have different currencies, or kinds of money. The currency of the United States is called **Federal Money.**

TABLE OF FEDERAL MONEY.

10 mills (m.) make	1 cent, . . .	c., ct.
10 cents,	1 dime, . .	d.
10 dimes,	1 dollar, . .	$
10 dollars,	1 eagle, . .	E.

158. All the denominations in this Table, except mills, are represented by coins. For convenience, other coins also have been issued. The coins of the United States are as follows:—

GOLD.	Double eagle,	worth	$20.
	Eagle,	"	$10.
	Half-eagle,	"	$ 5.
	Three-dollar piece,	"	$ 3.
	Quarter-eagle,	"	$ $2\frac{1}{2}$.
	Dollar,	"	$ 1.
SILVER.	Dollar,	"	$ 1.
	Half-dollar,	"	50 c.
	Quarter-dollar,	"	25 c.
	Dime,	"	10 c.
	Half-dime,	"	5 c.
	Three-cent piece,	"	3 c.
COPPER.	Two-cent piece,	"	2 c.
	Cent,	"	1 c.

157. What is said of different countries? What is the currency of the United States called? Recite the Table of federal money.—158. What denomination of this table is not represented by a coin? Name the gold coins, and their value. The silver coins. The copper coins.

MENTAL EXERCISES.

1. How many dollars are 5 double eagles worth?

Model.—1 double eagle is worth $20; and 5 are worth 5 times $20, or $100. Answer, $100.

2. How many dollars are 3 half-eagles worth? 4 quarter-eagles? 6 eagles? 2 double eagles?

3. How many cents in 2 half-dollars? In 4 quarter-dollars? In 7 dimes? In 3 half-dimes?

4. How many dollars in an eagle and a half-eagle? In 4 eagles and a half-eagle?

5. How many cents in 11 three-cent pieces? In 12 dimes and a half-dime?

6. How many eagles in 80 dollars?

Model.—$10 make 1 eagle; in $80 there are as many eagles as 10 is contained times in 80, or 8. Ans., 8 eagles.

7. How many dimes in 50 cents? In 6 half-dimes?

8. How many dollars in 18 half-dollars? In 24 quarter-dollars? In 70 dimes?

9. How many double eagles in $60? In $100?

Writing and Reading Federal Money.

159. In writing and reading federal money, the only denominations used are dollars, cents, and mills.

160. Dollars are denoted by this sign $, always placed before the number. They are separated from cents and mills by a point. The first two figures at the right of the point denote cents; the third figure, mills.

---

159. In writing and reading federal money, what denominations alone are used? 160. How are dollars denoted? Where are cents and mills found?

161. RULE.—*To write federal money, set down the dollars with a point at the right. Set the cents in the first two places at the right of the point, and the mills in the third place.*

If the cents are expressed by one figure, fill the vacant place with a naught. If there are mills, but no cents, fill both vacant places with naughts.

Ex.	Six dollars,	$6.
	Six dollars, fifty cents,	$6.50
	Six dollars, fifty cents, one mill,	$6.501
	Six dollars, five cents, one mill,	$6.051
	Six dollars, one mill,	$6.001

162. Write the following: let the points range in line.

1. Nine dollars, seventy cents.
2. Ninety dollars, five mills.
3. One hundred and forty dollars, seven cents.
4. Five dollars, seventy-five cents, seven mills.
5. Thirteen dollars, three cents, three mills.
6. Forty-one dollars, fourteen cents.

163. RULE.—*In federal money, read what is at the left of the point as dollars, the first two figures at the right of the point as cents, and the third figure as mills.*

164. Read the following:—

$400.276	$350.70	$54.
$112.009	$41.06	$6000.606
$907.072	$1011.004	$789.001

161. Recite the rule for writing federal money.—163. Recite the rule for reading federal money.

Addition of Federal Money.

165. Add \$72.25, $37\frac{1}{2}$ cents, \$9, and \$15.625.

\$72.25
.375
9.00
15.625
\$97.250

Here we are required to find the sum of several items in Federal Money. We must add things of the same kind. Therefore set dollars under dollars, cents under cents, &c., letting the points all range in line. Represent the half-cent as 5 mills. Then add in the usual way, and set off dollars in the result by placing a point under the points in the items added. Answer, \$97.25.

166. RULE.—*Write the several items, with their points ranging in line. Add, and place a point in the result under the points in the items added.*

NOTE.—As there are no mills coined, less than 5 mills in a result are disregarded in business dealings, and 5 mills or more are counted as an additional cent.

EXAMPLES FOR THE SLATE.

Read the following expressions: find their sum.

(1)	(2)	(3)	(4)
\$842.75	\$1269.454	\$327.00	\$27.50
34.03	73.401	562.009	4.516
460.983	1.011	437.09	.375
908.625	856.875	591.90	89.008
376.009	28.652	875.63	43.921

5. Bought a box of raisins for \$1.32, a bushel of apples for 88 cents, a cheese for \$5.94, and a barrel of sugar for \$27.62. What did the whole amount to?

165. Set down the given example. How must we place dollars, cents, &c.? How do we represent the half-cent? How do we set off dollars in the result? —166. Give the rule for the addition of federal money.

6. A farmer receives $15.37 for a cow, $75 for a horse, $3.13 for some potatoes, and $5.55 for some poultry. How much does he receive in all? *Ans.* $99.05.

7. Sold some velvet for $3.33, broadcloth for $18.75, silk for $12.50, muslin for $5.40, carpeting for $30.05, a shawl for $12.25. What is the amount of the bill? *Ans.* $82.28.

8. If a house cost $3487.75; repairs, $53.37; painting, $119.23; furniture, $1563.39; moving, $9; what was the whole cost? *Ans.* $5232.74.

9. A lady gives 25 cents for needles, $17.50 for a dress, $2.63 for trimmings, $1.50 for a cap, and 12 cents for thread. How much does she lay out? *Ans.* $22.

10. A man lends $68 to one friend, $443.75 to another, and $19.05 to a third. How much does he lend in all?

11. Add together sixty dollars; five cents, six mills; sixty cents; six hundred and fifty dollars, five mills; four mills; fifty-nine cents. *Ans.* $711.255.

12. There were taken up in a church collection, 16 cents, 3 three-cent pieces, 10 half-dimes, 8 dimes, 4 quarter-dollars, 3 half-dollars, and a two-dollar bill. What did the whole amount to? *Ans.* $6.05.

Subtraction of Federal Money.

167. If a person owing $143 pays $27.37, how much does he still owe?

$143.00
27.37
$115.63

We are here required to find the difference between $143 and $27.37. Set the subtrahend under the minuend, filling the vacant places of the latter with naughts. Place dollars under dollars, cents under cents, &c. Subtract in the usual way, and set off dollars in the remainder by placing a point under the other points. Answer, $115.63.

168. Rule.— *Write the subtrahend under the minuend, with their points ranging in line. Subtract, and place a point in the remainder under the other points.*

EXAMPLES FOR THE SLATE.

	(1)	(2)	(3)	(4)
From	$43.69	$101.467	$50.001	$64.
Take	17.748	88.35	9.099	.625

5. From forty-six dollars, two mills, subtract eighteen cents, nine mills. *Ans.* $45.813.

6. From one hundred dollars, three cents, take seven dollars, seven mills. *Ans.* $93.023.

7. A person, having bought goods to the amount of $65.76, gave the storekeeper a hundred-dollar bill. How much change must he receive? *Ans.* $34.24.

8. If a merchant sells goods that cost him $151.32 for $99.99, does he gain or lose, and how much?

9. Bought a cow for $37.25; paid on account $6.87. How much remains unpaid?

10. Paid for a lot $947.25; for erecting a house, $2345.47; for furniture, $1159. I sold the whole for $4500. Did I gain or lose, and how much? *Ans.* Gained $48.28.

11. A man worth $10000 gave away $956.38, and lost $1127.82. What was he then worth? *Ans.* $7915.80.

12. If a lady gives 12 cents for ink, 63 cents for pens, $13.30 for books, and $1.87 cents for paper, how much change must she get for a twenty-dollar bill? *Ans.* $4.08.

13. Bought $75 worth of hay, and $25.25 worth of corn; paid $49.88. How much is still due? *Ans.* $50.37.

14. From fifty dollars, take fifty cents, five mills.

---

168. Give the rule for the subtraction of federal money.

Multiplication of Federal Money.

169. What do 15 cows cost, at $16.345 each?

```
 $16.345
      15
 -------
   81725
  16345
 -------
$245.175
```

If 1 cow costs $16.345, 15 cows cost 15 times $16.345. Find the product, and point off from the right three figures for cents and mills, because there are three figures representing cents and mills in the multiplicand. Answer, $245.175.

170. RULE.—*Multiply in the usual way; point off from the right of the product, for cents and mills, as many figures as represent cents and mills in the multiplicand.*

EXAMPLES FOR THE SLATE.

	(1)	(2)	(3)	(4)
Multiply	$64.275	$15.89	$37.59	$293.872
By	9	12	13	56

5. What will 42 calves cost, at $3.75 apiece?

6. At 37½ cents apiece, what will 75 geese cost?

7. What will 890 cords of wood cost, at $3.78 a cord?

8. What will be the cost of 14½ yards of black silk, at $1.20 a yard? *Ans.* $17.40.

9. If a boy's wages are $4.75 a week, how much will he earn in a year, or 52 weeks? *Ans.* $247.

10. If a clerk earns $8 a week, and spends $4.75 a week, how much will he lay up in a year? *Ans.* $169.

11. What will it cost six persons to board for a year, at the rate of $5 apiece each week? *Ans.* $1560.

---

169. Go through the several steps in the example.—170. Recite the rule for the multiplication of federal money.

Division of Federal Money.

171. If 4 piano-fortes cost \$1501, how much do they cost apiece?

$$4)\underline{\$1501}$$
$$\$375\tfrac{1}{4}$$

If 4 pianos cost \$1501, one piano will cost one fourth of \$1501. Dividing, we find the quotient to be \$375¼.

$$4)\underline{\$1501.00}$$
$$\$375.25$$

If cents and mills are required in the answer, in stead of the fraction of a dollar, annex ciphers and continue the division. Point off in the product, for cents and mills, as many figures as represent cents and mills in the dividend. Ans., \$375.25.

25 cents make a quarter of a dollar. The answers agree.

172. Rule.—*Divide in the usual way; point off from the right of the quotient, for cents and mills, as many figures as represent cents and mills in the dividend.*

EXAMPLES FOR THE SLATE.

(1)	(2)	(3)	(4)
8) \$43.816	9) \$47.88	4) \$106	7) \$82.53
\$5.477	\$	\$	\$11.79

5. Divide \$12.48 equally among 12 persons.

6. If 36 hats cost \$88.92, how much is that apiece?

7. If a farmer sells 240 bushels of oats for \$132.48, how much does he get a bushel? *Ans.* \$.552.

8. Find $\frac{1}{104}$ of \$424.632. *Ans.* \$4.083.

9. Four partners bought some land for \$1150. They sold it for \$940 cash and \$500 worth of grain. How much did each make by the bargain? *Ans.* \$72.50.

---

171. Divide \$1501 by 4 in both the ways shown above.—172. Recite the rule for the division of federal money.

MISCELLANEOUS EXAMPLES IN FEDERAL MONEY.

1. If a person spends \$410.28 in a year, how much is that a week, allowing 52 weeks to a year? *Ans.* \$7.89.

2. A man buys 4 barrels of flour, at \$5.95 a barrel; 18 chickens, at 29 cents each; and 56 pounds of butter, at 27½ cents a pound. What does the whole cost? *Ans.* \$44.42.

Find the cost of each item; then add the three amounts.

3. Bought 3 pair of gloves, at \$.75 a pair; 12 yards of lace, at \$1.38 a yard; 10 yards of sheeting, at 45 cents a yard. What is the cost of the whole? *Ans.* \$23.31.

4. A lady buys 2 turkeys at \$1.25 each, and 5 bushels of potatoes at 94 cents a bushel. How much change will she receive for a \$20 bill? *Ans.* \$12.80.

5. A man pays for some land \$400 cash and \$192.80 in produce. If there were 57 acres, how much does the land cost him per acre? *Ans.* \$10.40.

6. If a boy buys 12 knives for \$7.50, and sells them at 75 cents apiece, how much does he make on each? *Ans.* \$.125.

Find the cost of 1 knife, and subtract it from the selling price.

7. Divide \$2117.71 equally among 35 families, and find the share of each. *Ans.* \$60.506.

8. Four persons contribute \$8000 for a speculation. The first puts in \$99.05; the second, \$2460.30; the third, \$986. What does the fourth put in? *Ans.* \$4454.65.

9. If 184 pounds of coffee are sold for \$52.44, what is the rate per pound? *Ans.* \$.285.

10. A father who has \$2450 in stock, a house valued at \$4750, and bonds to the amount of \$15040, divides the whole equally among his two sons and three daughters. What is the share of each? *Ans.* \$4448.

11. If 40 acres of meadow land are worth \$1260, what is $\frac{2}{3}$ of the tract worth? *Ans.* \$840.

REDUCTION.

173. How many cents in five dollars?

In 1 dollar there are 100 cents, and in 5 dollars 5 times 100 cents, or 500 cents. Answer, 500 cents.

We have here changed the denomination from dollars to cents, without changing the value. This process is called Reduction. We have *reduced* dollars to cents.

174. Reduction is the process of changing the denomination of a number without changing its value.

175. There are two kinds of Reduction:—

1. **Reduction Descending**, in which we change a higher denomination to a lower, as dollars to cents. Here we must multiply.

2. **Reduction Ascending**, in which we change a lower denomination to a higher, as cents to dollars. Here we must divide.

Reduction Descending.

176. Reduce $27 to mills.

```
$27
 100
-----
2700 c.
  10
-----
27000 m.
```

100 cents make $1; in $27, therefore, there are 100 times 27 cents, or (annexing two naughts) 2700 cents.

There are 10 mills in 1 cent; in 2700 cents, therefore, there are 10 times 2700 mills, or (annexing one naught) 27000 mills. Answer, 27000 mills.

---

173. Solve the example given. What have we done in this example?—174. What is Reduction?—175. How many kinds of reduction are there? What do we do in Reduction Descending? What operation must we perform? What do we do in Reduction Ascending? What operation must we perform?—176. Reduce $27 to mills, explaining the steps.

177. Reduce $27.465 to mills.

Reduce $27 to cents:	$27 \times 100 = 2700$ c.
Add in 46 cents:	$2700 + 46 = 2746$ c.
Reduce 2746 cents to mills:	$2746 \times 10 = 27460$ m.
Add in 5 mills:	$27460 + 5 = 27465$ m. *Ans.*

Compare this result with $27.465, the amount to be reduced. It is the same, with the dollar mark and point omitted.

178. General Rule for Reduction Descending.—*Multiply the highest given denomination by the number that it takes of the next lower to make one of this higher, and add in the number belonging to such lower denomination, if any be given.*

Go on thus with each denomination in turn, till the one required is reached.

179. *Federal Money.*—In federal money, the above rule is applied as follows. See the examples in §176, 177.

1. *To reduce dollars to mills, annex three naughts; to reduce dollars to cents, two; to reduce cents to mills, one.*

2. *To reduce dollars and cents to cents, or dollars, cents, and mills, to mills, simply remove the dollar mark and the point.*

180. Reduce the following:—

1. $624 to cents.	4. $450.63 to mills.
2. $125 to mills.	5. $29.172 to mills.
3. $.485 to mills.	6. $50000 to mills.

177. Reduce $27.465 to mills, explaining the steps. How does the result compare with the original amount?—178. Recite the general rule for reduction descending.—179. Give the rules for the reduction of federal money.

7. Reduce 5 eagles to mills. *Ans.* 50000 m.

8. How many cents in 28 eagles? *Ans.* 28000 c.

9. How many mills in 15 double eagles?

10. How many half-dimes in 73 eagles?

11. Reduce $450.59 to mills.

12. How many dimes are $67½ worth?

13. A lady gets a half-eagle changed to dimes. How many dimes should she receive? *Ans.* 50 d.

14. If a boy gets a $10 bill changed to half-dimes, how many should he receive? *Ans.* 200 half-dimes.

Reduction Ascending.

181. Reduce 27465 mills to dollars.

10 mills make 1 cent; therefore in 27465 mills there are as many cents as 10 is contained times in 27465. Dividing by 10 (cutting off one figure), we get 2746 cents, and 5 mills over.

10) 27465 m.
100) 2746 c., 5 m.
Ans. $27.465.

100 cents make 1 dollar; therefore in 2746 cents there are as many dollars as 100 is contained times in 2746. Dividing by 100 (cutting off two figures), we get $27, and 46 cents over. Answer, $27, 46 cents, 5 mills; or, $27.465.

Compare this result with 27465 mills, the amount to be reduced. We have simply pointed off three figures from the right, and inserted the dollar mark.

182. In § 177 we reduced $27.465, and obtained 27465 mills. In § 181 we reduced 27465 mills, and obtained $27.465. Thus it will be seen that Reduction Descending and Reduction Ascending prove each other.

---

181. Reduce 27465 mills to dollars, explaining the steps. How does the result compare with the amount given to be reduced?—182. By comparing the examples in § 177 and 181, what do we find?

183. General Rule for Reduction Ascending.—*Divide the given denomination by the number that it takes of it to make one of the next higher. Divide the quotient in the same way, and go on thus till the required denomination is reached. The last quotient and the different remainders form the answer.*

184. *Federal Money.*—In federal money, the above rule is applied as follows:—

To reduce mills to dollars, point off three figures from the right; to reduce cents to dollars, two; to reduce mills to cents, one.

EXAMPLES FOR THE SLATE.

1. Reduce 4790 mills to cents. *Ans.* 479 c.
2. Reduce 59195 mills to dollars.
3. Reduce 461063 cents to dollars.
4. How many dollars in 70000 mills?
5. How many dollars in 85310 cents?
6. How many eagles are 50 gold dollars worth?
7. Reduce 2500 dimes to eagles.
8. How many dimes are equal to 600 mills?
9. Reduce 46000 cents to double-eagles.
10. Reduce 4676 mills to cents.
11. How many half-eagles in 1500 cents?
12. How many dollars in 200 half-dimes?
13. How many double eagles in 1200 half-dimes?
14. Reduce 1623487 cents to dollars.
15. How many mills in 5 three-dollar pieces?
16. How many cents in 4 quarter-eagles?

---

183. Recite the general rule for reduction ascending.—184. Give the rule for the reduction of federal money.

COMPOUND NUMBERS.

185. A **Compound Number** is one consisting of different denominations: as, 3 dollars, 14 cents; 5 feet, 10 inches.

186. To show the relations that different denominations bear to each other, Tables are constructed. They must be learned perfectly.

English or Sterling Money.

187. The currency of Great Britain is called **English** or **Sterling Money.**

TABLE.

4 farthings (far., qr.),	1 penny, . .	d.
12 pence,	1 shilling, .	s.
20 shillings,	1 pound, . .	£.
21 shillings,	1 guinea, . .	guin.

188. The pound mark always precedes the number; as, £2. Farthings are sometimes written as the fraction of a penny; 2d. 3 far., or 2¾d.; 3d. 2 far., or 3½d.

The pound is simply a denomination. A gold coin called the Sovereign represents it. The Sovereign is worth $4.84. The English penny is worth about 2 of our cents.

Guineas, originally made of gold brought from Guinea, are no longer coined.

The Crown is a silver coin, worth 5 shillings.

---

185. What is a Compound Number?—187. What is the currency of Great Britain called? Recite the Table of Sterling Money.—188. How must the pound mark stand? How are farthings sometimes written? What represents the pound? What is the sovereign worth? The penny? What is said of guineas? What is the crown worth?

EXAMPLES FOR THE SLATE.

189. Recite the rules for Reduction, § 178, 183.

1. In £7 5s. 1 far. how many farthings?

Multiply the £7 by 20, to reduce them to shillings, because 20 shillings make a pound. Add in the 5 shillings.

Multiply 145s., thus obtained, by 12, to reduce them to pence, because 12 pence make a shilling. There are no pence in the given number to add in.

Multiply the 1740d., thus obtained, by 4, to reduce them to farthings, because 4 farthings make a penny. Add in the 1 farthing. Answer, 6961 far.

```
£7  5s.  1 far.
20
————
145 s.
 12
————
1740 d.
   4
————
6961 far.  Ans.
```

2. Reduce 15383 far. to pounds, shillings, &c.

```
  4 ) 15383 far.
     ————————
 12 ) 3845   3 far.
     ————————
2|0 ) 32|0   5d.
     ————————
     £16
Ans. £16 5¾d.
```

Divide by 4, to reduce to pence.

Divide the quotient, 3845d., by 12, to reduce it to shillings.

Divide the quotient, 320s., by 20, to reduce it to pounds. The quotient and remainders form the answer. Always mark the denominations throughout, as in these examples.

3. Reduce £75 to pence. *Ans.* 18000d.
4. Reduce 19s. 6d. to pence. *Ans.* 234d.
5. Reduce 15s. 3d. 2 far. to farthings. *Ans.* 734 far.
6. Reduce 8670d. to pounds, &c. *Ans.* £36 2s. 6d.
7. Reduce 16255s. to pounds. *Ans.* £812 15s.
8. Reduce 24681 far. to guineas, &c. *Ans.* 24 guin., &c.
9. Reduce £3 14s. $7\frac{1}{2}$d. to farthings. *Ans.* 3582 far.
10. Reduce 1920 far. to pounds, &c.
11. Reduce 3 guin. 10s. 6d. to farthings.
12. Reduce 16s. $10\frac{1}{4}$d. to farthings.
13. Reduce 1080 farthings to pounds, &c.
14. Reduce 8628 pence to pounds, &c.

Troy Weight.

190. Troy Weight is used in weighing gold, silver, precious stones, and liquors; also in philosophical experiments.

TABLE.

24 grains (gr.) make	1 pennyweight,	pwt.
20 pennyweights,	1 ounce, ...	oz.
12 ounces,	1 pound, . . .	lb.

EXAMPLES FOR THE SLATE.

1. Reduce 48494 gr. to pounds, &c. *Ans.* 8 lb. 5 oz. 14 gr.
   Divide by 24; then by 20; then by 12.
2. In 4 lb. 6 oz. 13 pwt. how many grains? *Ans.* 26232.
   Multiply 4 lb. by 12; add in 6. Multiply the result by 20; add in 13. Multiply this result by 24. Mark the denominations throughout.
3. In 100 lb. 1 gr. how many grains? *Ans.* 576001 gr.
4. Reduce 8976 pwt. to pounds. *Ans.* 37 lb. 4 oz. 16 pwt.
5. Reduce 9 oz. 5 pwt. 20 gr. to grains. *Ans.* 4460 gr.
6. How many pounds, &c., in 1180 oz.?
7. Reduce 16 lb. 10 oz. to pennyweights.
8. Reduce 5 lb. 5 pwt. to pennyweights.
9. Reduce 12 lb. 1 oz. 9 pwt. to grains. *Ans.* 69816 gr.
10. If a miner digs 20 pwt. of gold one day, and 37 the next, how many ounces will he have? *Ans.* 2 oz. 17 pwt.
11. How many pounds in two bars of silver, one of which weighs 240 pwt., and the other 360 pwt.?
12. How many pennyweights in 4 rubies, weighing 8 gr., 10 gr., 12 gr., and 24 gr.? *Ans.* 2 pwt. 6 gr.
13. A lady has 36 tablespoons weighing 34 pwt. each, and 24 teaspoons weighing 19 pwt. each. How many pounds do all her spoons weigh? *Ans.* 7 lb.

Apothecaries' Weight.

191. Apothecaries' Weight is used by apothecaries in mixing medicines. They buy and sell their drugs, in quantities, by Avoirdupois Weight, given on the next page.

TABLE.

20 grains (gr.) make	1 scruple,	sc. or ℈.
3 scruples,	1 dram, . .	dr. or ʒ.
8 drams,	1 ounce, . .	oz. or ℥.
12 ounces,	1 pound, . .	lb. or ℔.

192. The ounce and pound of Apothecaries' Weight are the same as in Troy Weight.

MENTAL EXERCISES.

1. How many ounces in $7\frac{1}{2}$ pounds of senna?

2. Reduce 1 ounce to grains.

3. Reduce $9\frac{1}{4}$ ounces to drams.

4. Which is greater, 15 ʒ or $1\frac{1}{2}$ ℥?

5. How many grains in 2 ʒ of magnesia?

6. How many scruples in $9\frac{1}{3}$ drams?

7. How many doses of 12 grains each in 1 ʒ of musk?

8. How many powders of ten grains each can a druggist make out of 2 ℈ of calomel?

9. Divide a dram of jalap into six powders; how many grains will there be in each?

10. If a druggist sells four customers 6 oz. of blue vitriol each, out of a five-pound package, how many pounds has he left?

11. How many grains in the following mixture: nitrate of silver, 5 gr.; opium, $\frac{1}{2}$ ʒ; camphor, 1 ℈?

Avoirdupois Weight.

193. Avoirdupois Weight is that in common use. By it are weighed all articles not named under Troy and Apothecaries' Weight; such as groceries, meat, coal, cotton, and all the metals except gold and silver.

TABLE.

16 drams (dr.) make	1 ounce, ...	oz.
16 ounces,	1 pound, ...	lb.
25 pounds,	1 quarter, ...	qr.
4 quarters,	1 hundred-weight,	cwt.
20 hundred-weight,	1 ton, ...	T.

194. The ounce of Avoirdupois Weight is less than the Troy ounce, but its pound is greater than the Troy pound.

195. It was formerly customary to allow 112 pounds to the hundred-weight, and 28 pounds to the quarter. But this is now seldom done, except in the case of coal, iron, and plaster bought in large quantities, and English goods passing through the Custom House.

Twenty hundred-weight of 112 pounds make a ton of 2240 pounds, which is distinguished as a Long Ton.

MISCELLANEOUS QUESTIONS.—In what denominations do British merchants keep their accounts? How much is a half-crown worth? What weight is used in weighing emeralds? In weighing hay? Coins? Cotton? In weighing drugs for a physician's prescription? Recite the Table used in philosophical experiments. Recite the Table used in weighing cheese. Recite the Table used in mixing drugs. Which is greater, the Troy pound or Apothecaries' pound? The Troy ounce or Avoirdupois ounce? The Apothecaries' pound or Avoirdupois pound? How many pounds were formerly allowed to the hundred-weight? In what alone is this now done? What is a Long Ton?

EXAMPLES FOR THE SLATE.

1. In 873450 drams, Avoirdupois Weight, how many tons, &c.? *Ans.* 1 T. 14 cwt. 11 lb. 14 oz. 10 dr.

Does this example fall under Reduction Descending or Reduction Ascending? Repeat the rule for Reduction Ascending. Name the numbers in order, by which we have to divide. In all examples in Reduction, be careful to mark the denominations throughout.

Prove this example by reducing your answer to drams. If the result agrees with the number of drams given above, your work is right.

2. Reduce 5 cwt. 21 lb. 4 oz. to ounces. *Ans.* 8340 oz.

Which kind of Reduction does this fall under? Repeat the rule for Reduction Descending. Name the numbers in order, by which we have to multiply. Why do we not first multiply by 20? As there are no quarters to add in, we may multiply the 5 cwt. by 100 at once, to reduce them to pounds, in stead of by 4 and 25.

3. Reduce 1 T. 1 cwt. 1 dr. to drams. *Ans.* 537601 dr.

4. Reduce 856702 drams to tons, hundred-weight, &c. *Ans.* 1 T. 13 cwt. 1 qr. 21 lb. 7 oz. 14 dr.

5. How many tons in 60000 lb. of lead? *Ans.* 30 T.

6. Reduce 3 cwt. 15 oz. to drams. *Ans.* 77040 dr.

7. How many pounds in 40¾ tons? *Ans.* 81500 lb.

8. How many ounces in 67⅓ cwt.?

9. Reduce 602000 oz. to tons. *Ans.* 18 T. 16 cwt. 1 qr.

10. What will 5½ cwt. of poultry cost, at 9c. a pound?

Reduce 5½ cwt. to lb.; then multiply by the price of 1 lb.

11. What cost 23¾ cwt. of beef, at 12c. a lb.? *Ans.* $285.

12. How many tons, &c., in 13 bales of cotton, weighing 550 lb. apiece? *Ans.* 3 T. 11 cwt. 2 qr.

13. How many pounds in 50 tons? In 50 long tons?

14. If I buy 370 long tons of coal, and sell 370 ordinary tons, how many pounds have I left? *Ans.* 88800 lb.

15. Bought 7 T. 18 cwt. of iron. What does it cost, at 4 cents a pound? *Ans.* $632.

16. How many ounce balls can be moulded out of 25 pounds of lead?

196. In connection with Avoirdupois Weight, learn the following

Miscellaneous Table.

14 pounds, . . .	1 stone of iron or lead.
56 pounds, . . .	1 firkin of butter.
100 pounds, . . .	1 quintal of dried fish.
196 pounds, . . .	1 barrel of flour.
200 pounds, . . .	1 bl. of beef, pork, or fish.

EXAMPLES FOR THE SLATE.

1. What is the cost of a firkin of butter, at 20 cents a pound? *Ans.* $11.20.

2. How many packages of 7 pounds each can be made out of a barrel of flour?

3. How many pounds in 42½ stone?

4. If a grocer who has 7 barrels of pork, sells half a barrel, how many pounds has he left?

5. How many firkins will 364 lb. of butter fill? *Ans.* 6½.

6. How many pounds in 17¼ quintals of codfish?

7. Bought 50 quintals of fish, at 6 cents a pound; what do they cost? *Ans.* $300.

8. If half a barrel of flour is sold for $2.94, what is the price per pound? *Ans.* 3c.

9. How many more pounds are there in 20 barrels of salted beef than in 20 barrels of flour?

10. If a barrel of pork brings $11.50, how much is that a pound? *Ans.* 5¾c.

11. How many ounces in 5 stone?

---

196. Recite the Miscellaneous Table. Is the pound here spoken of, the Avoirdupois or Troy pound? Is dried salted fish sold by the barrel or quintal? Is fish in pickle sold by the barrel or quintal?

Long Measure.

197. Long Measure is used in measuring length and distance. It begins with the inch.

1 inch.

TABLE.

12 inches (in.) make	1 foot, . . .	ft.
3 feet,	1 yard, . .	yd.
$5\frac{1}{2}$ yards,	1 rod, . . .	rd.
40 rods,	1 furlong, .	fur.
8 furlongs,	1 mile, . .	mi.

198. It is well to remember that (8×40) 320 rods or $(320 \times 5\frac{1}{2} \times 3)$ 5280 feet make a mile.

199. The Hand, used as a measure of the height of horses, is 4 inches. The Fathom, used as a measure of depths at sea, is 6 feet.

MENTAL EXERCISES.

1. How many inches will a man 6 feet high measure?
2. If a horse is 15 hands high, how many feet is that?
3. How many feet in 10 yards? In 2 rods?
4. Reduce 4 ft. 8 in. to inches.
5. How many inches in 2 fathoms?
6. How many yards in 6 rods? In 8 rods?
7. Reduce 108 inches to yards.
8. How many furlongs in $10\frac{5}{8}$ miles?
9. How many inches in a quarter of a yard?

---

197. In what is Long Measure used? Draw a line an inch long. Recite the Table.—198. How many rods make a mile? How many feet make a mile?—199. What is the Hand used in measuring? How many inches make a hand? What is the Fathom used in measuring? How many feet make a fathom?

EXAMPLES FOR THE SLATE.

1. Reduce 103 yards to rods.

$5\frac{1}{2}$ yards make a rod; hence we must divide 103 yards by $5\frac{1}{2}$, or $\frac{11}{2}$. To divide by $\frac{11}{2}$, multiply by the divisor inverted, $\frac{2}{11}$. Multiplying by the numerator 2, and dividing by the denominator 11, we get 18 rods, and 8 remainder.

```
     103 yds.
       2
11 ) 206 half-yd.
  rd. 18   8 hf.-yd.
Ans. 18 rd. 4 yd.
```

This remainder is not 8 yards, but 8 half-yards, since we reduced the original yards to half-yards when we multiplied by 2. Hence, to get the remainder in yards, divide it by 2. Answer, 18 rd. 4 yd.

In reducing yards to rods, then, if there is a remainder, divide it by 2, to bring it to yards.

2. Reduce 464 yd. to rods. *Ans.* 84 rd. 2 yd.

3. Reduce 1765 yd. to miles, &c. *Ans.* 1 mi. 5 yd.

4. Reduce 4355 in. to yards. *Ans.* 120 yd. 2 ft. 11 in.

5. Reduce 248 mi. to inches. *Ans.* 15713280 in.

6. How many miles, furlongs, &c., are there in 1051907 inches? *Ans.* 16 mi. 4 fur. 32 rd. 3 yd. 1 ft. 11 in.

7. How many inches are there in 10 mi. 1 fur. 29 rd. 3 yd. 2 ft. 10 in.? *Ans.* 647404 in.

8. Mt. Everest, a peak of the Himalayas, the highest mountain as yet surveyed, is 29002 feet above sea level. How many miles is this? *Ans.* 5 mi. 2602 ft.

200. In measuring drygoods, as cloth, muslin, lace, &c., the yard of long measure is used, divided into halves, quarters, and eighths.

1. What cost $19\frac{3}{4}$ yd. of lace, at 90c. a yard?

2. How many pieces a quarter of a yard long can be cut from 12 yards of silk?

3. If 2 dresses of $6\frac{7}{8}$ yd. and $10\frac{3}{8}$ yd. are cut from a piece of calico containing 40 yd., how much is left?

Square Measure.

201. Square Measure is used in measuring surfaces, which have length and breadth; such land, the sides of rooms, floors, &c.

202. A **Square** is a figure that has four equal sides perpendicular one to another—that is, leaning no more to one side than the other.

A Square Inch is a square whose sides are each an inch long. A Square Foot is a square whose sides are each a foot long.

A Square Inch.

1 inch.

1 inch. 1 inch.

1 inch.

Table.

144	square inches (sq. in.),	1 square foot,	sq. ft.
9	square feet,	1 square yard,	sq. yd.
$30\frac{1}{4}$	square yards,	1 square rod,	sq. rd.
40	square rods,	1 rood, ...	R.
4	roods,	1 acre, ...	A.
640	acres,	1 square mile,	sq. mi.

1. Reduce 40 A. 2 R. to square rods. *Ans.* 6480 sq. rd.
2. Reduce 14245 sq. rd. to acres. *Ans.* 89 A. 5 sq. rd.
3. Reduce 3 sq. mi. to square rods. *Ans.* 307200 sq. rd.
4. How many acres in 59 lots, of 2 roods each?
5. What will it cost to plaster four walls, each containing 270 sq. ft., at 20 cents a square yard? *Ans.* $24.
6. How many farms of 32 acres will 1 sq. mi. make?

---

201. In what is Square Measure used?—202. What is a Square? What is a Square Inch? A Square Foot? Recite the Table.

Cubic Measure.

203. Cubic Measure is used in measuring bodies, which have length, breadth, and depth or thickness; as stone, timber, earth.

204. A **Cube** is a body bounded by six equal squares.

A Cubic Inch is a cube, one inch long, one inch broad, and one inch thick. Each of its six sides is a square inch.

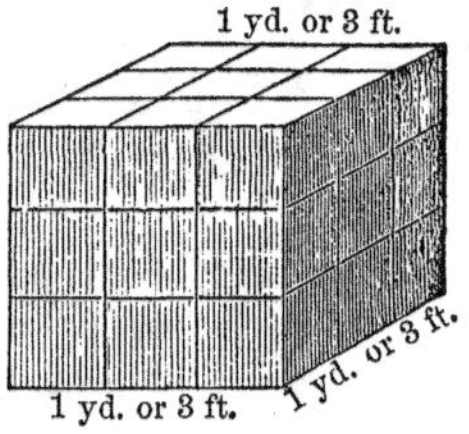

205. The engraving represents a Cubic Yard. It is 1 yard, or 3 feet, in length, breadth, and depth. It will be seen that each of its six sides is 1 square yard, or 9 square feet.

The top of this cube contains 9 square feet. Hence, if it were only 1 foot deep, it would contain 9 cubic feet. As it is 3 feet deep, it contains 3 times 9, or 27, cubic feet.

TABLE.

1728	cubic inches (cu. in.),	1 cubic foot,	cu. ft.
27	cubic feet,	1 cubic yard,	cu. yd.
40	cu. ft. of round or	1 ton or load,	T.
50	cu. ft. of hewn timber,		
16	cubic feet,	1 cord foot, . .	cd. ft.
8	cord feet,	1 cord, ...	Cd.

203. In what is Cubic Measure used?—204. What is a Cube? What is a Cubic Inch?—205. What does the engraving represent? Describe it. How many cubic feet does it contain? Recite the Table.

206. The ton in this Table is a measured ton; the Avoirdupois ton is a ton of weight. Round timber is wood in its natural state. A ton of round timber consists of as much as, when hewn, will make 40 cubic feet.

207. A cord of wood is a pile 8 feet long, 4 feet wide, and 4 feet high. Multiplying these dimensions together, we find 128 cubic feet in the cord. One foot in length of such a pile is called a cord foot.

208. Cubic Measure is often used in estimating the amount of work in solid masonry, in digging cellars, making embankments, &c.

EXAMPLES FOR THE SLATE.

1. Reduce 9 cubic yards, 16 cubic feet, 862 cubic inches to cubic inches. *Ans.* 448414 cu. in.

2. Reduce 746496 cu. in. to cubic yards. *Ans.* 16. cu. yd.

3. What will it cost to dig a cellar having a capacity of 4860 cu. ft., at 22 cents a cubic yard? *Ans.* $39.60.

   Find the number of cubic yards; multiply by the price.

4. What will it cost to make an embankment containing 108270 cu. ft. of earth, at 27c. a cubic yard?

5. Find the price of 1536 cd. ft. of wood, at $5 a cord.

6. How many cubic feet in 63⅛ cords?

7. Reduce 8 cu. yd. 469 cu. in. *Ans.* 373717 cu. in.

8. How many feet of hewn timber in 47⅗ tons?

Miscellaneous Questions.—What measure is used in finding the amount of surface in a floor? In estimating the amount of work in digging a well? In ascertaining the contents of a block of marble? In finding the distance between two places? In measuring one dimension, such as length? In measuring what has two dimensions, length and breadth? In measuring what has three dimensions, length, breadth, and thickness? How does the ton of Cubic Measure differ from the Avoirdupois ton? What is round timber? How much round timber makes a ton? What is meant by a cord of wood? A cord foot? In what is Cubic Measure often used?

Liquid Measure.

209. **Liquid** or **Wine Measure** is used in measuring liquids generally; as, liquors (beer sometimes excepted), water, oil, milk, &c.

TABLE.

4	gills (gi.) make	1 pint,	. . .	pt.
2	pints,	1 quart,	. .	qt.
4	quarts,	1 gallon,	. .	gal.
$31\frac{1}{2}$	gallons,	1 barrel,	. .	bar.
42	gallons,	1 tierce,	. .	tier.
2	barrels (63 gal.),	1 hogshead,		hhd.
2	hogsheads,	1 pipe,	. . .	pi.
2	pipes,	1 tun,	. . .	tun.

210. The wine gallon contains 231 cubic inches.

211. Liquids are put up in casks of different size, distinguished as barrels, tierces, hogsheads, pipes, and tuns; but these casks do not uniformly contain the number of gallons assigned to them in the Table, but only *about* that quantity. The contents are found by gauging, or actual measurement. When the barrel is used in connection with the capacity of cisterns, vats, &c., $31\frac{1}{2}$ gallons are meant; in Massachusetts, 32 gallons.

MENTAL EXERCISES.

1. How many gills in 2 qt.? In 1 gallon?
2. Reduce 3 qt. 1 pt. to gills.
3. How often can you fill a quart measure from a four-gallon tub? From a six-gallon tub?
4. What will a six-gallon can of milk cost, at 3c. a qt.?
5. If a milkman sells 10 qt. out of a four-gallon can, how many quarts remain?
6. How many gallons in 100 pints?

Beer Measure.

212. Beer Measure was formerly used in measuring beer and milk. It is still used by some for beer, though Wine Measure is fast taking its place.

TABLE.

2	pints (pt.) make	1 quart,	. . qt.
4	quarts,	1 gallon,	. . gal.
36	gallons,	1 barrel,	. . bar.
$1\frac{1}{2}$	barrels (54 gal.),	1 hogshead,	hhd.

213. The beer gallon contains 282 cubic inches. The gallon, quart, and pint of this measure, are therefore greater than those of Wine Measure.

EXAMPLES FOR THE SLATE.

1. Reduce 10000 gills to gallons. *Ans.* 312 gal. 2 qt.

2. In 16 hogsheads of wine, of 63 gallons each, how many gills? *Ans.* 32256 gi.

3. How many tuns will it take to hold 201600 gills of wine, allowing 252 gallons to the tun? *Ans.* 25 tuns.

4. How many quarts in 42 tierces of oil, of 42 gal. each?

5. Reduce 11 hhd. of beer to pints.

6. In 100000 pints of beer, how many gallons?

7. If a person buys a barrel of beer, containing 36 gallons, for $9, what does it cost him a quart?

8. How many pints in 18 gal. 3 qt. 1 pt.?

---

209. What is Liquid or Wine Measure used in measuring? Recite the Table.—210. How many cubic inches in a wine gallon?—211. What is said of the different denominations, barrels, tierces, &c.? How are the contents of a cask ascertained? How many gallons go to the barrel in Massachusetts? —212. For what was Beer Measure formerly used? For what is it still used? Recite the Table.—213. How many cubic inches in a beer gallon? Which is greater, the beer quart or the wine quart?

Dry Measure.

214. **Dry Measure** is used in measuring grain, seeds, vegetables, fruit, salt, coal, &c.

TABLE.

2 pints (pt.) make	1 quart, . . .	qt.
8 quarts,	1 peck, . . .	pk.
4 pecks,	1 bushel, . .	bu.
36 bushels,	1 chaldron, .	chal.

215. The quart of Dry Measure is greater than that of Liquid Measure.—What is called the Small Measure contains 2 qt.

216. Foreign coal is imported by the chaldron. American coal is bought and sold, in large quantities, by the ton; in small quantities, by the bushel.

EXAMPLES FOR THE SLATE.

1. Reduce 56 bu. 2 pk. 3 qt. to quarts. *Ans.* 1811 qt.
2. Reduce 8256 pt. to bushels. *Ans.* 129 bu.
3. How many bushels in 121⅔ chaldrons? *Ans.* 4380 bu.
4. Reduce 1597 qt. to bushels. *Ans.* 49 bu. 3 pk. 5 qt.
5. How many small measures in a bushel?
6. How many chaldrons in 1843 bushels of coal?
7. If a bushel of apples is bought for 80c., and retailed at 14c. a half-peck, what is the profit on them? *Ans.* 32c.
8. If 6 bushels of peaches are sold for $8.64, what do they bring a quart?

---

214. In what is Dry Measure used? Recite the Table.—215. How does the quart of Dry Measure compare with that of Liquid Measure?—216. What is imported by the chaldron? How is American coal bought and sold?

Time Measure.

217. The natural divisions of time are the year and the day. The year is the period in which the Earth makes one revolution round the Sun; the day, that in which it makes one revolution on its axis.

The year is divided into twelve calendar months, differing in length; the day, into hours, minutes, and seconds.

TABLE.

60 seconds (sec.) make	1 minute, . .	min.
60 minutes,	1 hour, . . .	h.
24 hours,	1 day, ...	da.
7 days,	1 week, . . .	wk.
12 calendar months or 365 days,	1 year, . . .	yr.
366 days,	1 leap year.	
100 years,	1 century,. .	cen.

218. The twelve calendar months (mo.), and the number of days in each, are as follows:—

	DAYS.		DAYS.
1st month, January,	31.	7th month, July,	31.
2d month, February,	28.	8th month, August,	31.
3d month, March,	31.	9th month, September,	30.
4th month, April,	30.	10th month, October,	31.
5th month, May,	31.	11th month, November,	30.
6th month, June,	30.	12th month, December,	31.

217. What are the natural divisions of time? What is the year? What is the day? How is the year divided? The day? Recite the Table.—218. Name the twelve calendar months in order, and the number of days in each.

219. The days in these months, added together, make 365 days in the year. Every fourth year (except three in four centuries) is a Leap Year; then February has 29 days, and the year 366.

220. The leap years are those that can be divided by 4 without a remainder; as, 1864, 1868, 1872, &c. But, of the even hundreds, only those that can be divided by 400 are leap years. The year 1900 will not be a leap year, but 2000 will be.

221. When we speak of a *month*, we mean a *calendar month*. The following lines will help the pupil to remember the number of days in each:—

"Thirty days hath September,
April, June, and November;
All the rest have thirty-one,
Except February alone;
Which has but four and twenty-four,
Till Leap Year gives it one day more."

EXAMPLES FOR THE SLATE.

1. In 30 days how many seconds? *Ans.* 2592000 sec.
2. Reduce 81920 min. to weeks. *Ans.* 8 wk. 21 h. 20 min.
3. Find the number of seconds in 13 wk. 3 da. 19 h. 25 min. 39 sec. *Ans.* 8191539 sec.
4. How many seconds in 4 successive years, three common years and one leap year? *Ans.* 126230400 sec.
5. Reduce 106847 sec. to hours. *Ans.* 29 h. 40 min. 47 sec.
6. How often will a clock, ticking once a second, tick in 24 hours?
7. When $12\frac{1}{2}$ hours of a day have gone, how many minutes remain?

---

219. How often does Leap Year occur? How many days in a leap year? Which month receives the additional day?—220. Which years are leap years?—221. Repeat the lines giving the number of days in the months.

Circular Measure.

222. Circular Measure is used chiefly in measuring angles and parts of circles, in determining latitude and longitude, and estimating the motions and positions of the heavenly bodies.

223. Every circle may be divided into 360 equal parts, called Degrees. The actual length of the degree will of course depend on the size of the circle. The Sign is a division of the circle used only in Astronomy.

TABLE.

60 seconds (″),	1 minute, . . .	′
60 minutes,	1 degree, . . .	°
30 degrees,	1 sign, ...	S.
12 signs (360°),	1 circle, ...	C.

Paper Measure.

24 sheets make	1 quire.
20 quires,	1 ream.
2 reams,	1 bundle.
5 bundles,	1 bale.

Collections of Units.

12 units make	1 dozen, doz.
12 dozen,	1 gross.
12 gross,	1 great gross.
20 units,	1 score.

---

222. In what is Circular Measure chiefly used?—223. Into what may every circle be divided? On what will the actual length of the degree depend? Recite the Tables.

EXAMPLES FOR THE SLATE.

1. Reduce 6 signs, 9 degrees, to degrees.

2. In 1000 minutes how many degrees?

3. Reduce 3 S. 18° to seconds. *Ans.* 388800″.

4. In 10000″ how many degrees, &c.? *Ans.* 2° 46′ 40″.

5. Reduce 45° 45′ 35″ to seconds. *Ans.* 164735″.

6. In 1000 bottles of porter how many dozen?

7. How many buttons in $18\frac{2}{3}$ dozen?

8. In 80064 tacks how many gross?

9. What will 480 bottles of ink cost, at $1 a dozen?

10. Pens are put up in boxes containing a gross; how many pens in 5 dozen boxes? *Ans.* 8640 pens.

11. If a box of pens is bought for 72 cents, what is the price of each pen? *Ans.* 5 mills.

12. If a paper of tacks containing a gross is sold for 6 cents, how many does that make for 1 cent?

13. How many great gross in 1378944?

14. How many sheets in one ream of paper?

15. How many sheets in $15\frac{5}{8}$ quires?

16. How many reams will 22480 sheets make?

17. How many sheets in 27 reams, 3 quires?

18. If half a ream of paper costs $2.40, what is the cost of each sheet? *Ans.* 1 ct.

19. If a stationer buys paper at $2.50 a ream, and retails it at 20c. a quire, what does he make on a ream?

20. How much paper will it take to make 1000 books, containing 6 sheets each? *Ans.* $12\frac{1}{2}$ reams.

21. If 2 circulars are printed on a sheet, how many can be printed on 5 bundles of paper? *Ans.* 9600 circulars.

22. How many reams of paper will it take to print 9696 such circulars? *Ans.* 10 reams, 2 quires.

MISCELLANEOUS EXAMPLES IN COMPOUND NUMBERS.

1. In 11959 grains, Apothecaries' Weight, how many pounds, ounces, &c.? *Ans.* 2 ℔. 7 ℥ 19 gr.

2. The highest point of the globe ever attained by man is the top of Mt. Chimborazo, 19699 feet above sea level. What is this height in miles? *Ans.* 3 mi. 3859 ft.

3. Reduce 1 square mile, 29 acres, 8 square rods, 7 square yards, to square feet. *Ans.* 29143881 sq. ft.

4. How many tumblerfuls, of half a pint each, will it take to fill a half-gallon pitcher?

5. Reduce $3\frac{1}{2}$ weeks to seconds. *Ans.* 2116800 sec.

6. In 1000 cd. ft. of wood, how many cords?

7. How many pages will there be in an edition of 2000 books, each book made of 12 sheets, and each sheet containing 24 pages? *Ans.* 576000 pages.

8. In 4 lb. 5 oz. 1 dr. 2 sc. 10 gr., how many grains are there? *Ans.* 25550 gr.

9. How many tons of hewn timber in 2500 cu. ft.?

10. Reduce 85274 pt. to bushels. *Ans.* 1332 bu. 1 pk. 5 qt.

11. How many lb. of silver will it take to make 4 dozen spoons weighing 18 pwt. each? *Ans.* 3 lb. 7 oz. 4 pwt.

12. What will 15 firkins of butter come to, at $23\frac{1}{2}$ cents a pound? *Ans.* $197.40.

13. How many doses of 6 drams each will 4 pounds of epsom salts make?

14. A farmer wishes to put up $367\frac{1}{2}$ bushels of potatoes in barrels holding 3 bu. 2 pk. each. How many barrels must he procure? *Ans.* 105 barrels.

How many pecks are to be put up? How many pecks will each barrel hold?

15. At 2 cents a dozen, how much will a great gross of buttons cost? *Ans.* $2.88.

16. How many days are there in the Spring months, March, April, and May?

17. How many days in the Summer months, June, July, and August?

18. How many days in the Autumn or Fall months, September, October, and November?

19. How many days in the Winter months, December, January, and February, when February falls in a leap year?

20. Reduce 19 cu. ft. to cubic inches. *Ans.* 32832 cu. in.

21. Reduce 5740 pwt. to pounds. *Ans.* 23 lb. 11 oz.

22. If from $\frac{3}{4}$ of an ounce of gold a jeweller takes enough to make 6 rings weighing $2\frac{1}{2}$ pwt. each, how many pennyweights will he have left?

23. The depth of water at a certain spot is found to be 31 fathoms, 3 ft. How many inches is this? *Ans.* 2268 in.

24. How many brushes, at 2s. each, can be bought for £8?

25. How long will 12 bushels of oats last a horse, if he is fed 8 quarts a day?

26. At the rate of $4 a cord, what is the value of a pile of wood, 8 feet long, 4 feet wide, and 4 feet high?

27. If sound moves at the rate of 1120 feet in a second, how many miles off is a cannon that is heard 11 seconds after it is discharged? *Ans.* 2 mi. 1760 ft.

28. If a family use 28 lb. of flour in a week, how long will 2 barrels last them?

29. How many pounds sterling will 6 dozen combs cost, at 9d. apiece? *Ans.* £2 14s.

30. If a locomotive goes a mile in 2 minutes, how many hours will it take to go 150 miles? *Ans.* 5 h.

31. Reduce 15 A. 3 R. 20 sq. rods, 15 sq. yards, 2 sq. ft. to square inches. *Ans.* 99597888 sq. in.

COMPOUND ADDITION.

224. Compound numbers may be added, subtracted, multiplied, and divided.

225. When compound numbers are added, the process is called **Compound Addition.** It combines addition and reduction ascending.

226. A person spends in one store £6 5s. 7d.; in another, £7 1s. 2d. 1 far.; in a third, £1 13s.; and in a fourth, £4 18s. 1d. 3 far. How much does he lay out altogether?

We are here required to find the sum of several compound numbers. We must add things of the same kind; therefore write numbers of the same denomination in the same column. Mark the denominations over the top.

£	s.	d.	far.
6	5	7	0
7	1	2	1
1	13	0	0
4	18	1	3
19	17	11	0

Beginning at the right, add the first column. Its sum is 4 farthings, which, by dividing by 4, we reduce to 1d. Set 0 in the column of farthings, and carry 1 to the next column.

The sum of the next column is 11. 11d. is not reducible to shillings, since it takes 12d. to make 1s. Set down 11, therefore, under the column of pence.

The sum of the shillings is 37s. = £1 17s. Set 17 under the column of shillings, and carry £1 to the next column.

The sum of the next column is £19. As pounds can not be reduced to any higher denomination, we set 19 at once under the column added. Answer, £19 17s. 11d. Hence the following rule:—

---

224. What operations may be performed on compound numbers?—225. When compound numbers are added, what is the process called? What operations does Compound Addition combine?—226. Explain the several steps in the given example.

227. Rule.—*To add compound numbers, set them down so that the same denominations may stand in the same column.*

Beginning at the right, add the denominations separately. Set each sum under the column added, unless it can be reduced to a higher denomination. If so, divide by the number that it takes to make one of that denomination; set the remainder under the column added, and carry the quotient.

Proof.—*Prove the addition, by adding in the opposite direction.*

EXAMPLES FOR THE SLATE.

Add the following compound numbers. Always mark the denominations over the top.

(1)

£	s.	d.
5	5	5
8	1	$7\frac{3}{4}$
2	0	$1\frac{1}{2}$
13	0	$11\frac{3}{4}$
6	6	6
34	14	8

(2)

lb.	oz.	pwt.	gr.
7	3	0	5
11	2	17	22
40	0	0	20
	6	18	16
2	10	15	17
61	11	13	8

(3)

℔	℥	ʒ	℈
2	11	6	2
10	8	3	1
14	10	2	2
	6	5	0
7	5	4	1
36	6	6	0

(4)

T.	cwt.	qr.	lb.	oz.	dr.
10	18	2	5	15	1
1	15	0	20	14	15
12	0	1	3	0	10
	13	0	7	1	11
2	2	2	20	7	8
27	9	3	7	7	13

(5)

mi.	fur.	rd.	yd.	ft.	in.
2	6	37	4	1	9
8	0	30	5	2	2
1	4	0	3	2	7
7	1	1	0	2	10
2	0	25	1	1	11
21	5	15	5	2	3

227. Recite the rule for Compound Addition. How may the addition be proved?

(6)

sq. yd.	sq. ft.	sq. in.
100	8	130
50	0	100
10	5	0
	8	143
13	2	8
175	7	93

(7)

cu. yd.	cu. ft.	cu. in.
4	26	1000
1	10	1541
	20	80
10	17	11
8	25	59
26	18	963

(8)

Cd.	cd. ft.
3	7
10	4
12	1
8	6
15	3
50	5

(9)

pi.	hhd.	gal.	qt.	pt.
1	1	30	3	1
	10	25	1	1
	25	0	2	0
	7	60	1	1
1	1	45	3	1

(10)

tun	pi.	hhd.	gal.	qt.	pt.	gi.
1	1	1	37	3	1	3
10	0	0	50	0	1	2
11	0	1	13	1	0	1
4	1	0	25	2	0	0
8	0	1	18	0	1	3

(11)

bu.	pk.	qt.	pt.
10	1	1	1
2	3	6	0
5	2	3	1
8	3	1	1
15	2	4	0

(12)

da.	h.	min.	sec.
15	18	50	49
1	13	59	59
4	23	47	2
	2	10	15
10	11	1	4

(13)

°	′	″
13	10	19
1	40	35
2	48	39
	30	40
10	45	45

14. Find the sum of 2 hhd. 50 gal. 3 qt. 1 pt. (Beer Measure); 10 hhd. 30 gal. 1 qt.; 11 hhd. 25 gal. 1 pt.; 25 hhd. 1 gal. 1 qt.; and 6 hhd. 52 gal. 3 qt. 1 pt.

Ans. 56 hhd. 52 gal. 1 qt. 1 pt.

15. Add together £7 13s. 3d.; £3 5s. 10d. 2 far.; £6 18s. 7d.; 2s. 5d. 3 far.; £4 3d.; and £17 15s. 4d. 2 far.

Ans. £39 15s. 9d. 3 far.

16. Add together 1 dr. 18 gr.; 2 dr. 1 sc. 15 gr.; 3 dr. 2 sc. 13 gr.; 4 dr.; and 6 dr. 1 sc. 7 gr. *Ans.* 2 oz. 2 dr. 13 gr.

17. A stable-keeper uses 8 bu. 2 pk. of oats, one day; 7 bu. 3 pk. 7 qt., the next; 7 bu. 6 qt., the third; 6 bu. 2 pk. 1 pt.,

the fourth; 8 bu., the fifth; 7 bu. 1 pk. 6 qt. 1 pt., the sixth; and 6 bu. 3 pk. 5 qt., the seventh. How much does he use during the week? *Ans.* 52 bu. 2 pk. 1 qt.

18. If one piece of cloth contains 40 yd.; another, $39\frac{7}{8}$ yd.; a third, $38\frac{3}{8}$ yd.; a fourth, $39\frac{1}{2}$ yd.; and a fifth, $40\frac{3}{4}$ yd.; how many yards are there in all? *Ans.* $198\frac{1}{2}$ yd.

19. What is the weight of four lots of iron, the first weighing 4 cwt. 3 qr. 20 lb.; the second, 2 T. 5 cwt. 14 lb.; the third, 1 T. 2 cwt. 2 qr.; and the fourth, 10 T. 19 cwt. 1 qr. 24 lb.? *Ans.* 14 T. 12 cwt. 8 lb.

20. How much paper will a printer use for three jobs, if the first job requires 6 bales, 1 bundle; the second, 4 bundles, 1 ream, 15 quires; and the third, 2 reams, 10 quires, 12 sheets? *Ans.* 7 bales, 2 bundles, 15 quires, 12 sheets.

21. A person owns five farms. The first contains 100 A. 1 R. 30 sq. rd.; the second, 600 A. 2 R. 10 sq. rd.; the third, 40 A. 1 R. 12 sq. rd.; the fourth, 250 A. 3 R. 2 sq. rd.; the fifth, 144 A. 20 sq. rd. How much land does he own in all?

Ans. 1136 A. 34 sq. rd.

22. A manufacturer makes four lots of pens. The first consists of 20 great gross, 7 gross, 5 dozen, and 6; the second, of 9 gross, 10 dozen, and 5; the third, of 15 great gross, 11 dozen; and the fourth, of 17 great gross, 3 gross. What is the whole amount made?

Ans. 53 great gross, 9 gross, 2 dozen, and 11.

23. Find the sum of 1 wk. 2 days 13 h. 40 min. 30 sec.; 2 wk. 6 days 10 h. 8 min. 3 sec.; 5 days 22 h. 55 min. 45 sec.; 4 h. 1 min. 15 sec.; and 1 wk. 2 days 4 h. 5 min.

Ans. 6 wk. 3 days 6 h. 50 min. 33 sec.

24. Add together 10 rd. 4 yd. 2 ft. 8 in.; 1 rd. 3 yd. 5 in.; 8 rd. 2 yd. 1 ft. 6 in.; 1 rd. 4 in.; and 2 yd. 1 ft. 9 in.

Ans. 22 rd. 2 yd. 8 in.

COMPOUND SUBTRACTION.

228. When one compound number is taken from another, the process is called **Compound Subtraction.** It combines subtraction and reduction descending.

229. A person who had 1 cwt. 8 lb. 15 oz. of cheese, sold 3 qr. 19 lb. 7 oz. How much had he left?

We are here required to find the difference between two compound numbers. Write the subtrahend under the minuend, placing numbers of the same denomination in the same column. Mark the denominations over the top.

	cwt.	qr.	lb.	oz.
	1	0	8	15
		3	19	7
Ans.			14	8

Begin to subtract at the right. 7 oz. from 15 oz. leave 8 oz.; set down 8 beneath, in the same column. 19 lb. can not be taken from 8 lb. We therefore take one of the next higher denomination, 1 qr., reduce it to pounds, and add it to 8 lb. 25+8=33; 19 lb. from 33 lb. leave 14 lb. Set down 14.

To balance the quarter added to the minuend, we must add 1 qr. to the subtrahend. This we do, by carrying 1 to the next column.

1 and 3 are 4. 4 qr. can not be taken from 0 qr. We therefore take 1 cwt., reduce it to quarters, and add it to 0 qr. 4 + 0 = 4 qr. 4 qr. from 4 qr. leave 0 qr. Carry 1. 1 from 1, 0. Answer, 14 lb. 8 oz. Hence the rule.

230. Rule.—*To subtract a compound number, set it under the minuend, placing numbers of the same denomination in the same column.*

---

228. When one compound number is taken from another, what is the process called? What does Compound Subtraction combine?—229. Go through the example, explaining the steps.—230. Give the rule for Compound Subtraction.

Beginning at the right, subtract each denomination separately, and place the remainder in the same column with the number subtracted.

If, in any denomination, the subtrahend is greater than the minuend, add to the latter as many as make one of the next higher denomination. Subtract, and carry 1 *to the next denomination of the subtrahend.*

PROOF.—*Add the remainder and subtrahend. If the sum is equal to the minuend, the work is right.*

EXAMPLES FOR THE SLATE.

(1)

	yr.	mo.	wk.	da.
From	17	8	3	1
Take	4	1	2	6
Ans.	13	7	0	2

(2)

T.	cwt.	qr.	lb.	oz.	dr.
13	18	1	20	0	13
10	0	3	21	12	0
3	17	1	23	4	13

(3)

℔	℥	ʒ	℈	gr.
24	7	2	1	16
16	10	3	2	17
7	8	6	1	19

(4)

mi.	fur.	rd.
60	0	0
40	7	39
19	0	1

(5)

A.	R.	sq rd.
69	3	25
10	0	38
59	2	27

(6)

cu.yd.	cu.ft.	cu.in.
144	12	123
89	23	869

(7)

ch.	bu.	pk.
30	10	1
10	8	3

(8)

tun	pi.	hhd.	gal.	qt.
10	1	1	50	1
1	0	0	60	3

9. A grocer buys 15 cwt. 20 lb. of sugar, and sells 10 cwt. 23 lb. How much remains unsold? *Ans.* 4 cwt. 3 qr. 22 lb.

10. From a piece of cloth containing $37\frac{3}{4}$ yd., have been cut off $6\frac{1}{4}$ yd. for one dress and $10\frac{1}{8}$ yd. for another. How many yards remain in the piece? *Ans.* $21\frac{3}{8}$ yd.

What is the proof in Compound Subtraction?

11. A person having £20 18s., spends £5 18s. 7½d. How much has he left? *Ans.* £14 19s. 4½d.

12. A farmer raises 100 bu. 3 pk. 2 qt. of wheat from one field, and 87 bu. 1 pk. 1 qt. 1 pt. from another. He sells 53 bu. to one person, and 37 bu. 2 pk. 1 qt. to another. How much has he left? *Ans.* 97 bu. 2 pk. 2 qt. 1 pt.

13. From a pile of wood containing 100 cords, I sold 10 Cd. 100 cu. ft. to one customer, and 18 Cd. 59 cu. ft. to another. How many cords remained? *Ans.* 70 Cd. 97 cu. ft.

14. The subtrahend is 19 mi. 7 fur. 8 rd. 2 ft. 10 in.; the minuend is 24 mi. 5 fur. 18 rd. 2 yd. 1 ft. 7 in. What is the remainder? *Ans.* 4 mi. 6 fur. 10 rd. 1 yd. 1 ft. 9 in.

15. A printer who has a bale of paper, uses 1 bundle, 1 ream, 5 quires, and 6 sheets. How much remains on hand?

16. A tailor uses 5 gross, 7 dozen, of buttons. How many has he on hand, if he had 1 great gross at first?

17. Poughkeepsie is 75 miles from New York. A man who starts to walk there, goes 22 mi. 3 fur. 15 rd. the first day, and 19 mi. 5 fur. 30 rd. the second. How much farther has he to go? *Ans.* 32 mi. 6 fur. 35 rd.

18. A jeweller, having a bar of silver weighing 2 lb. 6 oz., used 5 oz. 7 pwt. 12 gr. for one job, and 1 lb. 18 pwt. 7 gr. for another. How much was left? *Ans.* 11 oz. 14 pwt. 5 gr.

19. From a barrel of beer containing 54 gallons, a person drew 12 gal. 3 qt. one day, and 9 gal. 2 qt. 1 pt. another. How much was left?

20. From 39 sq. rd. 29 sq. yd. 128 sq. in., subtract 17 sq. rd. 16 sq. yd. 5 sq. ft.

21. A grocer has 1 cwt. 18 lb. of sugar in one barrel, 3 qr. 21 lb. in another, and 1 cwt. 2 qr. 11 lb. in a third. After selling 1 cwt. 3 qr. 15 lb., how much will he have left? *Ans.* 1 cwt. 3 qr. 10 lb.

COMPOUND MULTIPLICATION.

231. When a compound number is multiplied, the process is called **Compound Multiplication.** It combines multiplication and reduction ascending.

232. Multiply 4 bu. 3 pk. 7 qt. 1 pt. by 27.

27 being the product of 9 and 3, it is best to multiply by the factors in turn. Set 9, the first multiplier, under the lowest denomination of the multiplicand.

bu.	pk.	qt.	pt.
4	3	7	1
			9
44	3	3	1
			3
134	2	2	1

Begin to multiply at the right. 9 times 1 pint is 9 pt.; which, by dividing by 2, we reduce to 4 qt. 1 pt. Set 1 pt. in the column of pints, and carry 4 qt. to the next *product.*

9 times 7 qt. is 63 qt., and the 4 qt. carried make 67 qt., equal to 8 pk. 3 qt. Set 3 in the column of quarts, and carry 8 pk. to the next *product.*

9 times 3 pk. is 27 pk., and the 8 pk. carried make 35 pk., equal to 8 bu. 3 pk. Set 3 in the column of pecks, and carry 8 bu. to the next *product.*

9 times 4 bu. is 36 bu., and the 8 bu. carried make 44 bu. Set down 44.

Now multiply this product by 3, reducing and carrying in the same way. Answer, 134 bu. 2 pk. 2 qt. 1 pt.

233. Had we multiplied by 27 at once, we should have proceeded in the same way. As, however, we could not then have multiplied and divided in the mind, we should have had to write out the figures elsewhere, and set the results only under the line.—Multiply by 12 or less at once, in one line.

---

231. When a compound number is multiplied, what is the process called? What does Compound Multiplication combine?—232. Go through the example, explaining the steps.—233. Had we multiplied by 27 at once, how should we have proceeded? How must we multiply by 12 or less?

234. Rule.—*Set the multiplier under the lowest denomination of the multiplicand.*

Multiply each denomination in turn, and set the product under the number multiplied, unless it can be reduced to a higher denomination. If so, divide it by the number that it takes to make one of that denomination; set the remainder under the number multiplied, and carry the quotient to the next product.

Proof.—*When factors have been used in multiplying, multiply by the factors in reverse order. If the results agree, the work is right.*

EXAMPLES FOR THE SLATE.

	(1)			(2)			(3)			
	yd.	ft.	in.	£	s.	d.		oz.	pwt.	gr.
Multiply	1	0	9	10	10	10	8	7	5	2
By			4			3				7
Ans.	4	3	0	31	12	6	60	2	15	14

(4)					
T.	cwt.	qr.	lb.	oz.	dr.
	8	0	2	4	5
					6
2	8	0	13	9	14

(5)				
hhd.	gal.	qt.	pt.	gi.
1	2	3	1	2
				17
17	49	3	1	2

6. Multiply 5 cubic yards, 21 cubic feet, 643 cubic inches by 12. *Ans.* 69 cu. yd. 13 cu. ft. 804 cu. in.

7. How much cloth will it take for 7 suits of clothes, if each suit requires 7 yd. 3 qr.? *Ans.* 54 yd. 1 qr.

8. How much wood can a horse draw in 13 loads, if he draws 1 Cd. 1 cd. ft. each load?

---

234. Recite the rule for Compound Multiplication. What is the mode of proof, when factors have been used in multiplying?

9. How long will a man be in sawing 6 cords of wood, if he takes 7 h. 30 min. 45 sec. to saw 1 cord, allowing 10 working hours to each day? *Ans.* 4 days 5 h. 4 min. 30 sec.

Multiply in the usual way; then reduce the hours to working days by dividing by 10.

10. How much sugar is there in 21 hhd., each containing 11 cwt. 3 qr. 15 lb.? *Ans.* 12 T. 9 cwt. 3 qr. 15 lb.

11. Bought 15 yd. of broadcloth, at £1 3s. 6d. a yard, and 22 yd. of silk, at 7s. 8d. 2 far. a yd. What was the amount of the bill? *Ans.* £26 2s. 1d.

Find the cost of each item; then add.

12. The exact time in a year is 365 days 5 h. 48 min. $49\frac{7}{10}$ sec. What is the exact time in 50 years?

(50=5×10) *Ans.* 18262 days 2 h. 41 min. 25 sec.

13. How much brandy in 84 pi., each containing 128 gal. 2 qt. 1 pt. 3 gi.? *Ans.* 10812 gal. 1 qt. 1 pt.

14. If a man owning 5 farms, of 120 A. 1 R. 12 sq. rd. each, sells 450 A. 3 R. 25 sq. rd., how much land has he left? *Ans.* 150 A. 2 R. 35 sq. rd.

15. Bought 17 boxes of raisins, at 12s. 4d. a box; 5 bar. of flour, at £1 10s. 6d. a barrel; and 16 lb. of tea, at 5s. $3\frac{1}{4}$d. a lb. Paid on account £19 10s.; how much remains unpaid? *Ans.* £2 16s. 6d.

16. What will be the yield of 32 acres of wheat, at the rate of 24 bu. 2 pk. 7 qt. per acre? *Ans.* 791 bu.

17. If 2 gal. 2 qt. 1 pt. 1 gi. leak out of a water pipe in 1 hour, what will be the waste in 1 day? *Ans.* 63 gal. 3 qt.

18. Suppose a person to walk, on an average, 3 mi. 2 fur. every morning, and 3 mi. 20 rd. 1 yd. every afternoon; how far will he walk in two weeks? *Ans.* 88 mi. 3 fur. 2 rd. 3 yd.

19. If from 2 lb. of silver enough is taken to make a dozen spoons, weighing 1 oz. 10 pwt. 2 gr. each, how much will be left? *Ans.* 5 oz. 19 pwt.

COMPOUND DIVISION.

235. When a compound number is divided, the process is called **Compound Division.** It combines division and reduction descending.

236. Divide 148 gal. 3 qt. 1 pt. 3 gi. by 23.

Here we must use Long Division. Remember that a quotient is of the same denomination as the dividend from which it arises. Begin to divide at the left.

```
     gal.  qt. pt. gi.
23 ) 148   3   1   3 (6 gal.
     138
     ---
      10 gal.
       4
     ---
23 )  43 qt. (1 qt.
      23
     ---
      20 qt.
       2
     ---
23 )  41 pt. (1 pt.
      23
     ---
      18 pt.
       4
     ---
23 )  75 gi. (3 6/23 gi.
      69
     ---
       6 gi.
```

Ans. 6 gal. 1 qt. 1 pt. $3\frac{6}{23}$ gi.

Divide 148 gal. by 23: quotient, 6 gal.; remainder, 10 gal. Reduce the remainder to qt., and add in the 3 qt. in the dividend. 10×4=40 40+3=43.

Divide 43 qt. by 23: quotient, 1 qt.; remainder, 20 qt. Reduce the remainder to pt., and add in the 1 pt. in the dividend. 20×2=40 40+1=41.

Divide 41 pt. by 23: quotient, 1 pt.; remainder, 18 pt. Reduce the remainder to gi., and add in the 3 gi. in the dividend. 18×4=72 72+3=75.

Divide 75 gi. by 23: quotient, 3 gi.; remainder, 6 gi. As there is no lower denomination to reduce this remainder to, write it over the divisor in the form of a fraction, $\frac{6}{23}$.

Collect the several quotients for the answer.

If in any case there is no remainder, bring down the next denomination of the dividend, and proceed as above.

---

235. When a compound number is divided, what is the process called? What does Compound Division combine?—236. Go through the example, explaining the steps.

If the divisor is not contained in any dividend, set 0 in the quotient for that denomination, and reduce the dividend to the next.

237. Rule.—*Beginning at the left, divide each denomination in turn. When there is a remainder, reduce it to the next lower denomination, add in the number of that denomination in the given dividend, if any, and continue the division.*

When there is a remainder after the last division, place it over the divisor, in the form of a fraction, and annex it to the last quotient. The several quotients, each of the same denomination as its dividend, form the entire quotient.

Proof.—*Multiply the quotient by the divisor. If their product is equal to the dividend, the work is right.*

238. When the divisor is less than 12, use Short Division.

EXAMPLES FOR THE SLATE.

(1)

	mi.	fur.	rd.	yd.	ft.
7)	19	2	35	4	1
	2	6	5	0	$1\frac{6}{7}$

(2)

	cwt.	qr.	lb.	oz.	dr.
9)	27	3	17	13	9
	3	0	10	5	1

3. Divide 8 oz. 16 pwt. by 13. *Ans.* 13 pwt. $12\frac{12}{13}$ gr.
4. Divide 22 rd. 1 yd. 1 ft. 10 in. by 11. *Ans.* 2 rd. $5\frac{3}{11}$ in.
5. Divide £6 15s. 3d. by 10. *Ans.* 13s. 6d. $1\frac{1}{5}$ far.
6. Divide 86 bu. 1 pk. 1 pt. by 14. *Ans.* 6 bu. 5 qt. $\frac{5}{14}$ pt.
7. Divide 102 A. 1 R. 11 sq. rd. by 51. *Ans.* 2 A. 1 sq. rd.
8. Divide 4 gal. 2 qt. by 144. *Ans.* 1 gi.
9. Divide 40 cu. yd. 10 cu. ft. by 18.

237. Recite the rule for Compound Division. What is the proof?

10. If 31 clocks cost £113 13s. 4d., how much is that apiece? *Ans.* £3 13s. 4d.

11. If 6 oz. 7 dr. (Apothecaries' Weight) of magnesia are put up in 60 equal parcels, how much will each weigh?

Ans. 2 sc. 15 gr.

12. A silversmith makes seven teapots, of equal weight, out of 9 lb. 1 oz. 14 pwt. 5 gr. of silver. What is the weight of each? *Ans.* 1 lb. 3 oz. 13 pwt. 11 gr.

13. If 47 casks, of the same size, hold 1686 gall. 1 pt. of beer, how much will each contain? *Ans.* 35 gal. 3 qt. 1 pt.

14. If a traveller goes 600 miles in 1 day 7 h. 35 min. 20 sec., what is his average time per mile? *Ans.* 3 min. $9\frac{8}{15}$ sec.

15. Divide 6 bales, 3 bundles, 1 ream, of paper into 8 equal parts. *Ans.* 4 bundles, 7 quires, 12 sheets.

16. A farmer puts up 1000 bushels of apples in 350 barrels of uniform size. How many bushels, &c., does each barrel contain? *Ans.* 2 bu. 3 pk. $3\frac{3}{7}$ qt.

17. An estate worth £2570 is to be divided as follows: the widow is to have one third of the whole, and the rest is to be divided equally between seven children. What is the widow's share, and what each child's?

Ans. { Widow's, £856 13s. 4d.
Child's, £244 15s. 2d. $3\frac{3}{7}$ far.

18. What is the weight of 13 crowns, each weighing 18 pwt. $4\frac{4}{11}$ gr.; 14 shillings, each weighing 3 pwt. $15\frac{3}{11}$ gr.; and 9 sixpences, each weighing 1 pwt. $19\frac{7}{11}$ gr.?

Ans. 1 lb. 3 oz. 3 pwt. $15\frac{3}{11}$ gr.

19. A farmer having 450 bu. 1 pk. 1 qt. of corn, after selling 425 bu. 3 pk. 6 qt., distributed the rest equally among 5 poor families. How much did each receive?

Ans. 4 bu. 3 pk. 3 qt. $1\frac{3}{5}$ pt.

20. Divide 182° 5′ 12″ by 12.

MISCELLANEOUS EXAMPLES.

1. A person has $741.35 in one bank, $350 in another, and $1129.88 in a third; how much has he in bank altogether?

2. At 40 cents a yard, what will be the cost of 5 miles of telegraph cable? *Ans.* $3520.

3. The silk-worm's thread is about $\frac{1}{3000}$ of an inch thick; the spider's web is about $\frac{1}{6}$ as thick. How thick is it?

4. Reduce $\frac{27}{63}$ to its lowest terms.

5. How many pens in $8\frac{3}{16}$ gross?

6. Three boys, gathering nuts, agree to divide equally all they get. The first gathers 540, the second 960, the third 720. How many does each receive? *Ans.* 740 nuts.

7. Cut from a piece of cloth containing $36\frac{1}{8}$ yd., enough cloth to make 10 coats, each requiring $2\frac{1}{4}$ yd., and how much will be left?

8. The product of two factors is $29\frac{4}{9}$. One of the factors is $8\frac{5}{6}$; what is the other? *Ans.* $3\frac{1}{3}$.

9. If 1 sovereign is worth $4.84, what are 27 sovereigns worth?

10. A man who has $\frac{1}{2}$ mile the start of another, gains on him 30 rods more. How many feet is he then ahead of the other?

11. If a boy who has been in the habit of sleeping 9 h., rises an hour earlier every day, how many days will he save in 5 years, allowing for one leap year? *Ans.* 76 da. 2 h.

12. From $2\frac{1}{4}$ subtract $1\frac{3}{8}$, and multiply the remainder by $3\frac{1}{7}$. *Ans.* $2\frac{3}{4}$.

13. If a man has 3 small farms, of 8 fields each, and each field contains 2 A. 3 R. 22 sq. rd., how much land has he in all? *Ans.* 69 A. 1 R. 8 sq. rd.

14. From a piece of cloth containing $121\frac{1}{2}$ yd., a tailor made 18 coats, which took one third of the whole piece. How many yards did each coat contain? *Ans.* $2\frac{1}{4}$ yd.

15. How many pints in $26\frac{1}{3}$ gal.? *Ans.* $210\frac{2}{3}$ pt.

16. A man worth $10000, at his decease, left $225.75 to the poor, and the rést to his five sons, in equal shares. How much did each get? *Ans.* $1954.85.

17. How many five-acre lots are there in a square mile?

18. A trader, making four speculations, in the first gains $1520, in the second loses $460, in the third loses $720.50, and in the fourth gains $986.25. Does he gain or lose on the whole, and how much? *Ans.* Gains $1325.75.

19. A man owning a mine, sells $\frac{1}{6}$ of it to one person, $\frac{2}{7}$ of it to another, and $\frac{2}{5}$ of it to a third. How much does he sell altogether? *Ans.* $\frac{179}{210}$.

20. How many lots of half a rood each can be laid out in $6\frac{1}{2}$ acres? *Ans.* 52 lots.

21. How many guineas are equivalent to £42?

22. A man buys 7 head of cattle for $210. Two die, and he sells the rest for $34.50 apiece. How much does he lose?

23. Add $\frac{1}{6}$ of 90, $\frac{2}{3}$ of 129, and $\frac{5}{7}$ of 1260. *Ans.* 1001.

24. If from 1 gal. of wine 3 qt. $1\frac{7}{8}$ pt. are sold, how much will remain? *Ans.* $\frac{1}{8}$ pt.

25. How many sheets of paper in three lots, consisting of $1\frac{1}{2}$ reams, 5 quires, and 1 bundle? *Ans.* 1800 sheets.

26. Three persons, going on an excursion, agree to share the expense equally. They hire a carriage for $2.25, a boat for 75c. Their dinner costs them 50c. each, and their tea $1.20 for all four. How much has each to pay? *Ans.* $1.90.

27. Which contains the most units, $4\frac{1}{2}$ dozen, $\frac{3}{8}$ of a gross, or $\frac{1}{36}$ of a great gross?

28. From $\frac{4}{5}$ of $6\frac{1}{9}$ subtract $\frac{1}{3}$ of $2\frac{2}{3}$. *Ans.* 4.

29. How much is a man worth, over and above his debts, whose whole property is valued at £5100, and who owes one person £75 10s. and another £427 19s. 6d. 2 far.?

Ans. £4596 10s. 5d. 2 far.

30. When a man who has $3\frac{1}{2}$ mi. to walk, has gone half-way, how many miles has he yet to go?

31. How many dozen in $\frac{3}{4}$ of a great gross?

32. How many times is $8\frac{1}{9}$ contained in $9\frac{1}{8}$? *Ans.* $1\frac{1}{8}$.

33. A crown is worth 5s. How many crowns are equivalent to £25 10s.?

34. Two regiments, of 1125 men each, reach a river. They have 5 boats, capable of carrying 25 men each. How many trips must the boats make, to get them over?

35. Reduce to a simple fraction $\frac{2}{3}$ of $\frac{4}{9}$ of $\frac{1}{2}$ of $\frac{27}{6}$ of $\frac{5}{7}$ of $\frac{3}{2}$ of $1\frac{1}{6}$. *Ans.* $\frac{5}{6}$.

36. An officer, pursuing a thief, gains on him $\frac{4}{5}$ of a mile every hour. If the thief is $7\frac{3}{5}$ miles ahead, how many hours will it be before he is overtaken? *Ans.* $9\frac{1}{2}$ h.

37. How many more seconds does a leap year contain than an ordinary year? *Ans.* 86400 sec.

38. A lad who has 288 marbles, loses $\frac{1}{6}$ of them, sells $\frac{1}{4}$, and gives away $\frac{5}{12}$ of them. How many marbles has he left? *Ans.* 48 marbles.

39. How many cows, at $30 apiece, must a farmer give for 6 acres of land, at $15 an acre? *Ans.* 3 cows.

Find the value of the land; how many cows, at $30, will pay for it?

40. A miller buys $22\frac{1}{4}$ cords of wood at $4 a cord, and pays for it in flour at $6 a barrel. How many barrels must he give?

41. A weekly paper has been in existence $10\frac{3}{4}$ years. If each paper contains 8 pages, and each page 6 columns, how many columns has it contained during that time?

www.ingramcontent.com/pod-product-compliance
Lightning Source LLC
LaVergne TN
LVHW021410110826
845150LV00007B/1865